THE DIGITAL WATCH STORY

ANDREW MORTEN

AMBERLEY

About the Author

Andrew Morten has had an interest in electronics from an early age and went on to gain his degree in Electrical Engineering from Coventry in 1987. He spent his career working on various projects in the industrial and defence sectors, at companies such as Plessey, Racal and General Electric. He is now retired and lives in Leicestershire.

First published 2025

Amberley Publishing
The Hill, Stroud,
Gloucestershire, GL5 4EP

www.amberley-books.com

ISBN: 978 1 3981 2006 8 (print)
ISBN: 978 1 3981 2007 5 (ebook)

British Library Cataloguing in Publication Data.
A catalogue record for this book is available from the British Library.

Typeset in 10pt on 13pt Celeste.
Typesetting by SJmagic DESIGN SERVICES, India.
Printed in the UK.

Appointed GPSR EU Representative: Easy Access System Europe Oü, 16879218
Address: Mustamäe tee 50, 10621, Tallinn, Estonia
Contact Details: gpsr.requests@easproject.com, +358 40 500 3575

Contents

Introduction

If you wear an electronic watch today, it may be an all-singing, all-dancing smartwatch or fitness tracker, complete with a touch-sensitive colour screen. You may use it for relaying messages from your phone, paying for your drink at a coffee shop or even telling the time. Or perhaps it's a Casio with a built-in calculator, or maybe it's a ruggedised model made for extreme sports. Whatever your electronic watch of choice, its ancestors can be traced all the way back to the early 1970s.

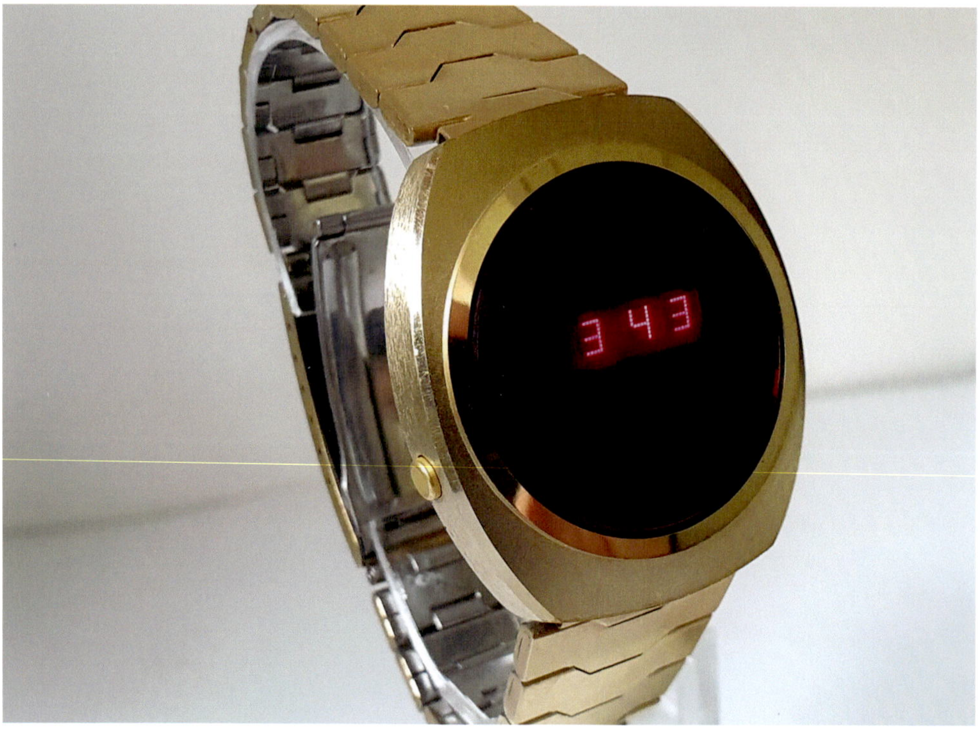

An early LED digital watch from US company Benrus.

The story of the electronic watch begins in the era of the space race and the Apollo moon landings. The technology from those programs gave us the microchip, and the start of a new age of miniaturisation. Pundits at the time were forecasting all manner of uses for this new technology, such as automation in the home, robots, computers and even flying cars. But one of the first commercial uses for the microchip was the digital watch. However, with an initial price tag of over $20,000 in today's money, the first watches were out of reach for all but the very rich. Indeed, most people's first glimpse of these new marvels was on the wrists of movie stars such as Roger Moore and Peter Sellers, or celebrities such as Elton John and Joe Frazier. In 1972, they really were seen as a miracle of technological progress. With their glowing red numerals, the wrist-mounted computer looked like something *Star Trek's* Captain Kirk would wear, and really captured the public's imagination in the age of space exploration.

No sooner had the 'basic' digital watch been perfected, manufacturers raced to make them smaller and cheaper. The power-hungry LED displays used in the early watches were soon replaced with the black-on-white of liquid-crystal displays. These new LCD watches didn't require a button press to see the time – it was always visible. Imagine that! And viewing the time in bright sunlight was no longer a problem either. Best of all, the electronics consumed so little power that the expensive batteries would now last for a year instead of just weeks.

A solar-powered LCD watch from Casio.

Next, other features were added, and a watch with a calculator went on sale in 1975. Seiko was behind many of the innovations in digital watches from the mid-1970s such as the first watch with a chronometer. Other world-firsts from Seiko included a TV watch, a voice recorder and a watch with a diary.

The latest evolution of the digital watch is of course the 'smartwatch', companion to the smartphone. These have many uses, from delivering the weather forecast to recording your blood pressure whilst out for a jog. In the fifty years since the first electronic watch, they have been transformed from a humble timepiece into, for some anyway, an indispensable piece of wrist technology. This book looks at some of the more notable watches produced on that journey and the companies behind them.

A modern smartwatch.

1

Light-Emitting Diode (LED) Watches

All time-keeping devices require a component that vibrates or oscillates regularly to measure time. In a mechanical watch, this is the balance wheel, which is driven back and forth by the spring, providing a regular 'tick'. The very first mechanical timepieces appeared around 500 years ago in the form of a pocket watch, but the more familiar wristwatch has been around since after the First World War. Although most people came to own one, they had several drawbacks. Having to remember to wind them up before they stopped was a chore, but overwinding could cause damage. Also, being precision instruments, they were quite expensive, and the very best mechanical watch accuracy was only around one minute a week.

There were a couple of attempts to improve on this, firstly by powering the balance wheel by an electric motor, and then using a tuning fork arrangement, resonating at 300 Hz. This achieved an improved accuracy of around half a minute per month, but neither of these were successful, and so research began into a third type of electronic watch.

It was known that a quartz crystal could be made to vibrate at its natural resonant frequency when incorporated into an electronic circuit. The smaller the crystal and the higher the frequency, the better the accuracy would be in a watch. With a frequency of 32,768 Hz, accuracy of around a minute a year was possible.

The first watch to use a quartz crystal was developed by watch manufacturer Seiko, who launched their Quartz Astron model in 1969. It looked just like a conventional watch with its analogue hands, but was accurate to five seconds a month, a hundred-fold improvement over any other watch of the time.

This was a big step towards the all-digital watch, but the equally challenging part was that of converting the quartz crystal ticks into a digital time display. This was a job for the microchip.

In a typical early digital watch, the microchip counts exactly 32,768 pulses from its quartz clock and registers one second. The logic circuitry in the watch's chip then adds this to the running total of hours, minutes and seconds that has passed since the time was set. The final step is to convert the time stored within the chip into a display of the time in numerals, which is where the Light-Emitting Diode (LED) comes in.

The phenomenon of producing light from a diode was discovered in 1907 by H. J. Round of Marconi Labs. He discovered a physical effect where at the junction between

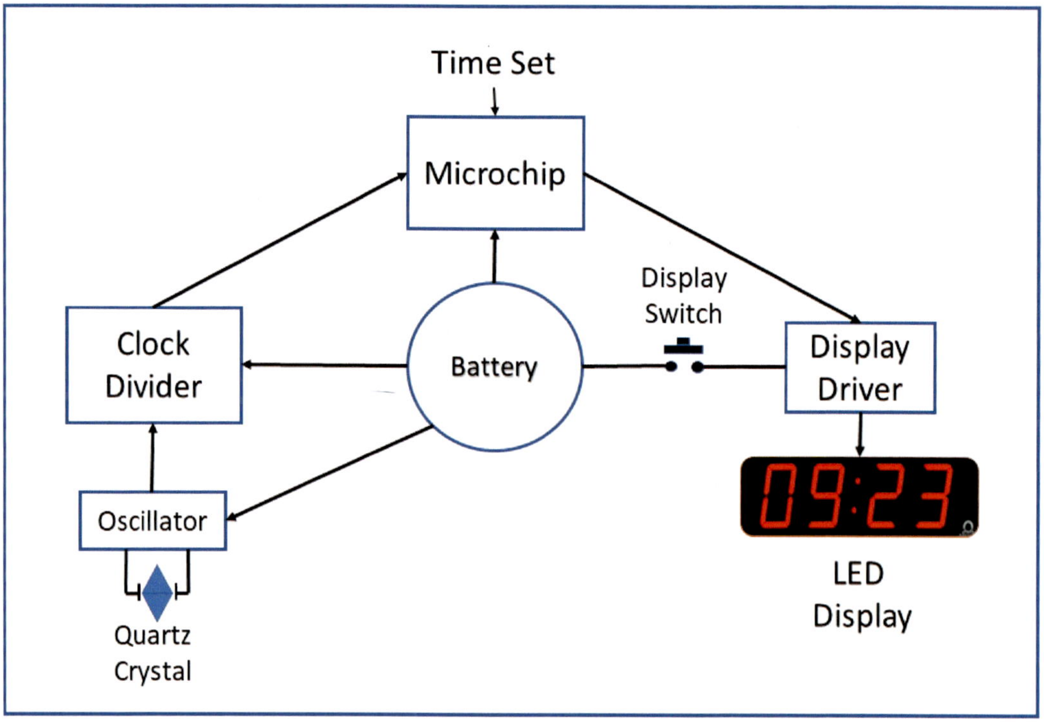

Simplified circuit diagram for a digital watch.

two different materials, light was emitted when a voltage was applied. At the junction, electrons and positive ions in the material combine and release a photon of light in the process. In the early 1960s, practical LEDs were made using semiconductors – a special material with directional conductivity. The earliest LEDs emitted infrared and were, for example, used in TV remote controls. Later, red visible light was possible and these were soon incorporated into modules called seven-segment displays. Here, seven bar-shaped LEDs are arranged as a figure eight which can display all the numbers from zero to nine just by switching on the appropriate LED segments. Four LED modules can then be used to show hours and minutes in digital format.

This was quite a complex job for the electronics at the time and early chips contained over 3,000 transistors, squeezed onto a silicon chip just a few millimetres square. To extend the short battery life of early watches, the display was normally off and would light for around one second only when a button was pressed.

To pack all of the electronics, wiring and batteries into a module measuring around 1 inch diameter by a quarter inch deep was, some thought, an impossible task. But the commercial benefits from developing a completely new type of timepiece were motivation enough. The photo opposite shows both sides of an early watch module. Although the chip itself is hidden inside, the four LED modules for the display can be seen, each of which contains the separate LED segments to create the numbers.

The maker of the first digital watch was the Hamilton Company in the US, who developed the Pulsar range. Other American manufacturers of the time were Texas Instruments and

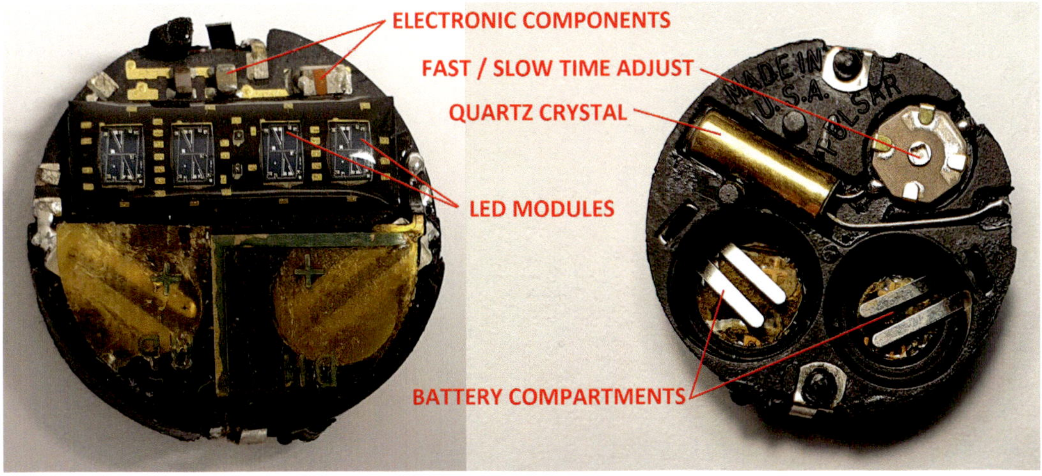

Front and rear views of an early LED watch module.

watch companies Bulova and Timex. This chapter explores some of the most important and innovative watches produced in the heyday of the LED watch – the 1970s.

Pulsar P1 (1970)

The story of the first digital watch starts in 1966, when two engineers working for the Hamilton Company in Lancaster, Pennsylvania, were commissioned by Stanley Kubrick to work on his 1968 film *2001: A Space Odyssey*. The film required the design of a futuristic timepiece and engineers Richard S. Walton and John M. Bergey created a digital clock based on 'Nixie' tubes – a type of glass valve containing wire-shaped numbers that could be made to glow. Sadly, the Odyssey Clock never actually made it into the film, but the experience made Walton and Bergey believe that a digital watch was possible, and they convinced Hamilton to set up an Electronic Watch Project Division within the company.

At the same time, a company called Electro/Data in Garland, Texas, was also working on designs for a solid-state watch using LED displays. Talks between the two companies led to an agreement being signed in 1969 to jointly develop a completely new type of watch – silent, accurate and with no moving parts.

The technical challenge was huge at the time. It required fitting a computer into a watch-sized space, and nothing remotely like it had been done before. The watch would use LED technology from NASA and a design containing nearly 4,000 transistors and other components that would need to be hand soldered onto tiny ceramic circuit boards, all squeezed into the watch case. A screen of synthetic ruby would light up on a button press and display the time in a digital format. The design even contained a light sensor to dim the display in the dark.

By the time that the first prototype was ready in 1970, the watch had a name: Pulsar, after the rotating stars that emit beams of radiation at ultra-precise frequencies. And the Electronic Watch Project became Hamilton's Pulsar Division.

The first digital watch – the Pulsar (Limited Edition) P1.

The Pulsar was unveiled to the general public on *The Tonight Show* on 5 May 1970. When the show's host, Johnny Carson, was told that the price tag would be in the thousands of dollars, he quipped: 'The watch will tell you the exact moment that you went bankrupt.' Although Johnny Carson didn't seem impressed, the rest of America was. The *New York Times* called it a 'Space-age wrist computer' and a watch expert called the display 'the greatest technological leap since the watch spring in 1675'. However, it took a further two years to create a production version. Electro/Data had problems producing the modules, and because of the delays Pulsar started work to design their own modules. But in April 1972, Electro/Data delivered and the first generation of Pulsars went on sale. The solid gold version retailed for $2,100 (over $20,000 in today's money) at high-end retailers like Tiffany's and the 400 units sold out in three days.

Pulsar thought that there would be no technical problems and so they sealed the case-back, leaving only access for two battery ports. Unfortunately, many of the watches started to fail due to a flaw in one of the Electro/Data chips, and all of the watches sold had to be recalled. Luckily, the more reliable Pulsar modules were now available, which saved the day for Pulsar's reputation. It also ended the agreement with Electro/Data.

Interestingly, the Hamilton Company had a watch-making division, but this was in no way involved in designing or manufacturing the Pulsar. A restructuring in 1971 saw Hamilton become HMW (Hamilton-Metals-Wallace) and in 1972 a subsidiary called TCI (Time Computer Inc) was formed for the Pulsar brand and the watches became known from then onwards as 'Time Computers', rather than wristwatches.

The first watches were only available in 18K gold. (Courtesy D. L. Klein)

Pulsar P2 (1973)

A new, improved version known as the P2 followed and was priced more affordably. There was a gold-plated model, known as 'gold filled', for $375 and a steel case model at $295. For those with deeper pockets, there was a 14K solid gold case version for $1,275 and an 18K solid case and bracelet version for $2,100. The P2 was a more rounded shape than its predecessor, and the single button required to display the time was moved from the bottom of the case to the right-hand side. The watch was heavily advertised by HMW at the time, promoting its reliability, accuracy and lack of any moving parts to wear out. A unique feature of Pulsars was the way the time was set. Placing a magnet near positions on the watch's case back allowed the hours and minutes to be adjusted. The small magnet was stored in the clasp when not in use. The LED segments making up the digits in early watches such as these were actually tiny points of light when viewed closely.

The Pulsar P2 became the must-have watch for the rich and famous of the day. Keith Richards, Jack Nicholson, Elvis Presley and even the Emperor Haile Selassie of Ethiopia were all seen wearing one. In the opening scenes of that year's Bond film – *Live and*

The first completely new way to tell time in 500 years...

invented and made in U.S.A.

- Pulsar® is a solid-state Time Computer® no larger than a wristwatch.
- It has no moving parts to wear out.
- It never needs maintenance, oiling, or cleaning.
- There are no dials, hands, gears, springs, tuning forks, or motors; nothing to wind up or run down.
- Accurate to within one minute a year.
- Performance and accuracy unconditionally guaranteed for 3 years.*
- Very modestly priced from $275.

■ Pulsar, the world's first solid-state Time Computer for your wrist, is a remarkably accurate, virtually trouble-free new way to tell time.

Its brain is a high frequency quartz crystal that splits a second into 32,768 parts. As a result, we can guarantee that Pulsar will gain or lose no more than one minute a year. (Timing will be adjusted to this tolerance, if necessary.)

Quartz is only a small part of the wonder of Pulsar.

As long as the case and time screen remain intact, Pulsar is water resistant up to a depth of 100 feet.

It didn't miss a second in tests when subjected to shocks and vibrations up to 2500 times the force of gravity. High magnetic fields won't permanently damage it.

When you press the command button, the exact time flashes on the ruby-red time screen and stays on for 1.25 seconds. Continue to press the button and the seconds flash on, one after another.

Pulsar is powered by two energy cells that will last about a year if you check the time an average of 25 times a day.

Pulsar is available in stainless steel at $275; in a 14 kt. gold-filled case and bracelet at $375.

You may well have to wait for the Pulsar of your choice. Production cannot be hurried. Each Pulsar is individually tested before it is released for sale.

Inspect Pulsar at the nearest fine store. It will give you a new pride in American craftsmanship and technology.

Unprecedented 3-year guarantee

The performance and accuracy of the Pulsar Time Computer module is unconditionally guaranteed for three years from date of original purchase.

In the unlikely event that the Computer malfunctions within this guarantee period, your Pulsar jeweler will replace the entire module on the spot, free of charge.

If your Pulsar jeweler does not have a replacement module in stock, he will send your Pulsar to our service laboratory. We will repair or replace the module and send it back to you within 48 hours from the time of receipt.

(*Guarantee does not cover energy cells and does not apply if module has been damaged by abuse or accident.*)

ACTUAL SIZE

Pulsar

When you touch the button, Pulsar tells you the time

12:08

Pulsar

PULSAR®

THE TIME COMPUTER®

Subsidiary of HMW Industries, Inc.

For FREE literature write: PULSAR Box 1609, Lancaster, Pa. 17604

In Canada: Henry Birks & Sons, Ltd.

Advert for the Pulsar P2.

Gold-plated P2 model seen in the film *Live and Let Die.*

Let Die – James is seen pressing the side button of his P2 to reveal the time in glowing numbers, furthering its desirability as the cool 1970s gadget. By the end of the year tens of thousands were being sold every month, making it the fastest-selling high-end watch in the world.

Pulsar P3 (1973)

A P3 model followed in late 1973 – the first LED watch with the date. Known as the Date Command, the new left-hand button was used to momentarily display the month and date, which was correct for a non-leap year, i.e. twenty-eight, thirty or thirty-one days. The right-hand button briefly displayed the hours and minutes, and a further press displayed the seconds. The time was set in the same way as the P2 – by placing a magnet in the HR or MIN slots on the back of the watch. The same slots were used to set the date when the date button was held in.

Early watches had lightweight bracelets by Champion USA in steel or gold plate, sometimes known as 'lobster style' from the way they were segmented. Later watches had solid-steel link bracelets. The magnet that fitted into the bracelet for time setting was also upgraded from the steel type used in the P2 to a black ferrite magnet. Inside, the module used was essentially the same as used in the P2, with a few extra components and logic for the date. Its success meant that it was produced right up until 1975. Incidentally, the

Pulsar P3 with lightweight hollow-segment bracelet.

original-sized batteries are no longer available, but modern-sized batteries can be fitted with spacer rings.

Pulsar P4 (1975)

In 1975, the last major revision – the P4 – was released. It was quite similar in appearance to the P3, but the buttons, which were left and right on previous Pulsar models, were now at the six and twelve o'clock positions. It also came with a smaller setting magnet inside the clasp. Pictured in the advert opposite are two different P4 designs: a gold-plated Executive and a squarer-faced stainless-steel design known as the Men's Dress model. Inside, the module was of a completely new design, thanks to improvements in watch technology. The LED segments had been composed of tiny dots, but the P4 had solid bars, which made the display brighter. Another improvement was the Auto-Command feature, which allowed the user to activate the display with just a flick of the wrist. This was achieved by incorporating a tilt switch, consisting of a small tube with two contacts and a drop of mercury. A sudden movement would shift the mercury and make the circuit, lighting the display. There was also a 'Classic' version issued in mid-1975, which went back to the case style and button positions of the old P2 but had the new module fitted.

Long after the novelty wears off, the pleasure of owning a Pulsar® grows

* * *

The Time Computer® that started the stampede to solid state digital time, is a possession to be proud of for a lifetime...or two.

No one really needs a timepiece as fine as Pulsar, possibly the first heirloom to be created by modern computer technology.

This small miracle from Lancaster, Pa. is so incredibly accurate that a list of world-wide time stations has to be packed with every Pulsar to assure you of a source of time signals precise enough to check its accuracy.

Long-lasting excellence to the smallest detail

The power cells, in the models pictured, are designed to last at least a year if you check the time an average of 25 times a day.

The little lamps that flash the time and date should retain 80% of their original brightness after 100 years of continuous use. Impact 2500 times the force of gravity has not harmed Pulsar in tests. It is completely water resistant to a depth of at least 100 feet as long as case and time screen remain intact.

Full 3-year warranty

Every Pulsar is continuously tested for at least one week before it is released for sale.

And every Pulsar carries a *full* 3-year warranty from date of original purchase. Complete details are packed with every Pulsar. Ask your authorized Pulsar jeweler to show them to you before you buy.

If you take a quiet satisfaction in owning the acknowledged leader and in the impeccable service offered by the finest jewelers, purchasing a Pulsar will reward you well.

For more information, write Time Computer, Inc., Lancaster, Pa. 17604; or call, toll-free (800) 233-0262. In Canada: Henry Birks & Sons Ltd. Overseas: Pulsar s. a., Box 339 Stand, 1211 Geneve 11, Switzerland.

Pulsar®
THE TIME COMPUTER
Sold and serviced throughout the free world

New . . . above: slim and shining in stainless steel, Model 3408, $295. Below: flashes time at a flick of the wrist . . . 14 kt. gold-filled case with matching bracelet, Model 5201, $395. Shown actual size.

Time Computer Inc. advert for the Pulsar P4.

Ladies' Pulsars

Although the early Pulsar models were not explicitly targeted at men, purchasers were almost exclusively male as the watches were large and heavy due to the size of the electronics and batteries. But from 1974, further advances in miniaturisation of the electronics allowed for smaller versions of Pulsars to be marketed towards women.

This ladies' Pulsar dates from late 1974/early 1975 and was known as the 'cushion' model due to the display shape. It was essentially a smaller version of the Pulsar P4 with its gold-plated case and bracelet. It featured AM/PM indication, a 'smart' calendar with automatic adjustment for twenty-eight-, thirty- and thirty-one-month days, a magnet to adjust the time and a light sensor.

In 1976, Pulsar created their Touch Command range, so called because they did away with the buttons. Instead, there were two contacts at the twelve o'clock and six o'clock positions, which merely had to be touched to display the time or date. Using skin

Ladies' 14K gold-filled Pulsar model.

Ladies' Pulsar Touch Command model.

conductivity, contact with a finger was enough to make the circuit and light the display. This cushion-shaped model featured an AM/PM indicator, date with smart calendar, water-resistance to 30 m and a light sensor to brighten the display in daylight.

By 1975, Pulsar were manufacturing around 120,000 watches a year, and this would peak in 1976 with around 150,000 units. But all good things come to an end and Time Computer Inc.'s best days were behind it. A combination of new LCD technology and a huge inventory of unsold stock led to TCI's demise. Time Computer Inc. sold their registered brands, Pulsar & Time Computer, to Rhapsody Inc. of Philadelphia on 1 July 1977. TCI also made many other models, including some with emerald-green displays, and these were hastily assembled from parts and transferred to Rhapsody as part of the agreement.

For the next year, Rhapsody continued to supply the existing worldwide dealer network with Pulsars. On the back cover of the last and final Pulsar brochure, Rhapsody was acknowledged as the owner of the Pulsar and Time Computer brands. Although Rhapsody could have continued to manufacture Pulsar watches, in September of 1978 they sold the brands to the Seiko Corporation of Japan. As of today, the only trademark maintained by Seiko is Pulsar. The diagram below shows the various companies involved in the creation and evolution of the Pulsar brand.

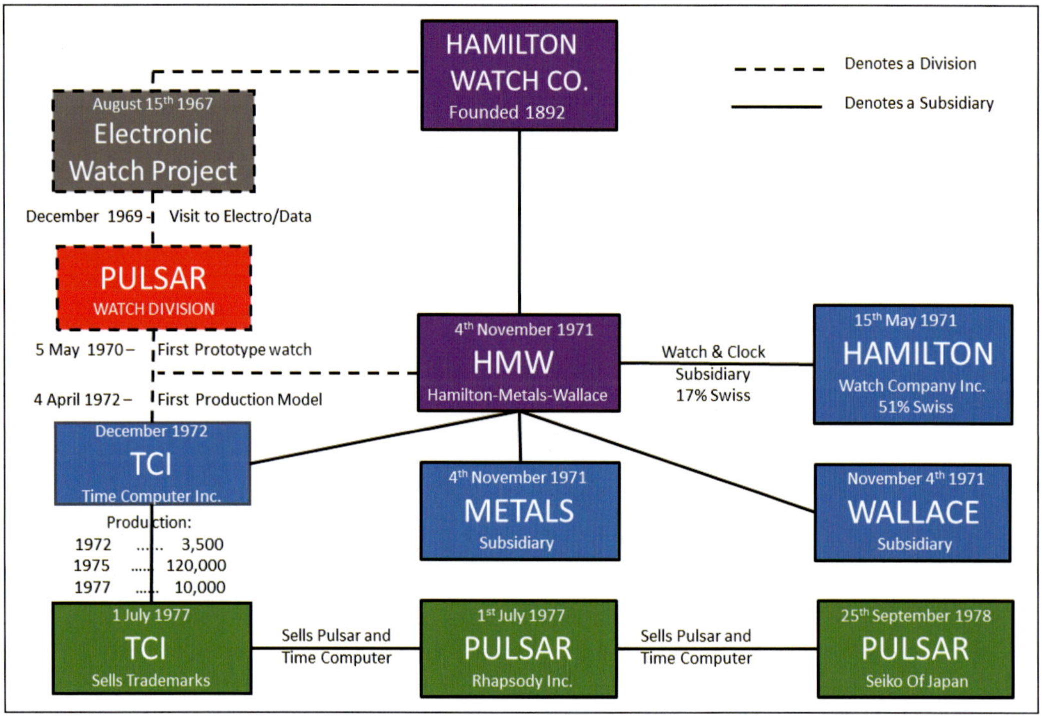

The companies involved with the Pulsar brand.

Time Computer's Pulsar watches were the US's greatest success story in the area of watches. Today, they are very collectable; early working P2 models with their box and papers can reach up to £1,000. One of the original P1 18K gold models sold on eBay several years ago for over $17,000.

Hamilton QED (1973)

At the end of 1971, when restructuring of the Hamilton parent company saw its name change to HMW, Swiss company SSIH purchased a 17 per cent stake of the newly split-off Hamilton Watch Co., which in 1974 became 51 per cent, giving the Swiss ownership of what was once America's greatest watchmaker.

Swiss-Hamilton then went on to market their own design watches called the QED or QTC, based on watch modules made by Pulsar. QED stands for 'Quartz Electronic Digital' whereas QTC means 'Quartz Time Computer'. Early watches (before Hamilton was controlled by the Swiss) are labelled QTC, indicating the trademark was owned by Time Computer Inc. After Hamilton was restructured and acquired by the Swiss, they changed the branding to QED to remove any association of Hamilton's watches with Pulsar TC Inc., as that was still an American business. However, all Hamilton LEDs are Pulsars with the Hamilton logo and everything inside is identical, with either P2 modules fitted to one-button versions or P3 electronics in two-button models.

Right: Hamilton gold-plated QED model.

Below: Hamilton watches used Pulsar modules.

Omega Time Computer (1974)

Everyone with more than a passing interest in watches will have heard of Omega and know that it is one of the world's more prestigious watch marques. Today, the Omega brand, along with Tissot, are owned by Swiss Group SSIH, but back in 1973 they entered the digital watch market with the Time Computer 1 or TC1. However, Omega didn't make the case or the module but added their logo to bought-in parts and sold the first ever European LED watch.

Omega commissioned the Star Watch Case Company to make the cases that would accommodate Pulsar P2 modules. The cases therefore have the identical backs, locking rings, batteries, magnet-setting feature and single-button positions as their Pulsar cousins. Genuine Omega modules are stamped with the Omega logo inside, but many that have died will have been resurrected by replacing the module with one from a cheaper P2. The TC1 was available in solid gold and yellow or white gold-plate.

In 1974, the TC2 was released. This retained the same case as the TC1 but used the Pulsar P3 module and so had the extra button on the left-hand side for the date. As with the TC1, a module original to the watch will have the Omega logo inside. Cases were available in stainless steel or gold plate. Today, a TC1 or TC2 could be purchased for £700–£2,500 depending on the model, condition and its box.

Omega's TC2 model.

Two-button model has a date function.

Compu Chron (1973)

Compu Chron was a Japanese watch company that produced some of the earliest digital watches from 1973. The watch modules, however, were manufactured by the American Hughes Aircraft Company, founded in February 1934 by Howard Hughes. Hughes Aircraft were an aerospace and defence company and a pioneer of the electronics industry. During the US space program, they made the Surveyor craft for Moon surveys in the 1960s and the Galileo craft for the mission to Jupiter in the 1980s. But in the early 1970s they produced digital watch modules, containing their own microchips, which were used in many LED watches of the time, including Compu Chron.

This rather large gold-cased watch is from mid-1973 and housed what was known as the 29-mm Hughes Aircraft module. This module was unusual in that it used a high-frequency quartz crystal oscillating at 786.432 kHz, which was twenty-four times the normal frequency. The idea was that the higher frequency would be more stable and not require a trimming device inside the watch for adjustment. However, this idea proved false, and aging and temperature still caused the time to drift and so trimmers were added very soon after. Early models had the point-type LED segments and showed time, date and seconds. Later models had bar-type segments and extra functions such as leap year and day of the week.

Left: Gold-plated Compu Chron model.

Below: Stainless-steel Compu Chron from 1976.

Another more common version of the Compu Chron was this stainless-steel model manufactured from 1976 to 1978. They were fitted with the later 25-mm Hughes Aircraft modules, which have the display button top right and the setting button bottom left. In 1978, a company called Unisonic was using the Compu Chron brand and had built up a very large stock of cases. Unfortunately, Hughes stopped production of their 25-mm modules and so Unisonic were left with thousands of cases they could not sell. Years later they were purchased and ended up on eBay, where they sold for pennies!

Wittnauer Polara (1974)

Most people will have heard of the Swiss watchmaker Longines, probably best known for supplying the timekeeping for many sporting events such as skiing, horse racing and the Commonwealth games. The company was founded in 1832 and specialised in accurate time measurement. They developed the first pocket watch chronograph in 1878 and the first stopwatch with one-hundredth-second precision in 1916. Their logo of the winged hourglass is the oldest trademark still in use today. In 1950 Longines bought the US watch company Wittnauer, founded by Albert Wittnauer, who emigrated from Switzerland to America in 1872. Along with watches, the Wittnauer company specialised

A Polara 1 is in the British Museum's collection.

Wittnauer in its presentation box.

in aviation; Amelia Earhart's aircraft was fitted with Wittnauer instruments. In the Second World War, Wittnauer supplied their compasses to the US army and air force. The new Longines-Wittnauer company kept the two brands separate.

This high-end model was produced in 1974 and its price tag of $165 was a considerable amount in its day. However, with its 35-mm-wide watch face and 14-mm depth, it was not to everyone's taste. This stainless version was named the Polara 1, and there is one in the British Museum as an example of an 1970s LED watch. There was also a 14K gold-plated option called the Polara 2. The Polara face was made of mineral glass and the movement was made by the Hughes Aircraft Company.

The Wittnauer company retained its reputation for most of the twentieth century, and was bought by Bulova in September 2001. They continue to make watches to this day.

Bulova Accuquartz (1974)

The Bulova Watch Company was founded in the US in 1875 by immigrant Joseph Bulova and established its operations in New York. Since 2008 it has been owned by Japanese conglomerate Citizen Watch. In 1926 Bulova had the first advertisement broadcast on radio, announcing the first beep in history: 'At the tone, it's eight o'clock, Bulova Watch Time.' Another milestone came in 1941 with the world's first television advertisement. Before a baseball game between the Brooklyn Dodgers and Philadelphia Phillies, New York station WNBC broadcast a clock-shaped testcard where the second hand swept round for a minute showing the Bulova logo.

As well as mechanical watches, Bulova produced a range of electronic watches called Accutron, which pre-dated the digital watch. They were based around a tuning fork powered by a one-transistor circuit oscillating at 360 Hz. Produced for a short while from 1960, they had an accuracy of one minute per month and also made a slight hum.

Large case gave the AccuQuartz the nickname of 'bigblock'.

The first LED watch from Bulova in 1974 was the AccuQuartz, also known as 'bigblock' because of its large, solid case. Bulova didn't do their own R&D, but bought parts from other manufacturers. The module for the AccuQuartz was a modified version of a very common module at the time made by Hughes Aircraft, used in many watches between 1973 and 1977.

The unusual thing about this watch though is that rather than using buttons to operate it, the AccuQuartz used a 'crown', similar to the winder found on mechanical watches. This meant that this solid-state watch had some moving parts! This was possibly because customers changing up from a mechanical watch would find it more familiar to operate.

The crown did three things: pressing it in displayed the time, so acted like a button; pushing it in further displayed seconds; and pulling the crown outwards displayed the date. The crown was connected to a 'stem' inside the module which had to align very accurately with three sets of contacts. This made them unreliable and any watch that developed a fault would likely be due to a failure of the crown mechanism.

Bulova Computron N6 (1976)

Probably the most striking watch Bulova ever produced was the Computron N6. Not only did it have a futuristic name and shiny gold case reminiscent of C-3PO from *Star Wars*, but having the display on the side of the case gave it a unique look. Known as the 'Trapizoid', 'Sideview' or just 'Wedge', this design feature was perhaps to make it easier to see when driving. With both hands on the wheel, the display would be facing you. It was therefore

The Computron N6 with unusual side-on display.

also known as the 'driver's watch', not to be confused with a 'diver's watch'! But this seems a minor advantage as one hand would still have to be taken off the wheel to press the display button. It is more likely that the side display made it easier to read in direct sunlight. Reading the time outdoors was a problem for LED watches with their relatively dim displays, and one or two other manufacturers tried a similar design.

The Bulova's cool shape though makes them one of the most collectible LED watches. Bulova thought so too and so in 2019 they re-launched a modern version of Computron for around £250. Bulova have kept the futuristic 1970s design but updated the electronics to give it a long battery life. As well as gold, chrome or black are available, and with either red or blue displays.

Trafalgar (1973)

Most digital watch manufacturers bought in parts from other companies and added their own branding. One of the British companies that took this route was Trafalgar, with entrepreneur and Chairman Willy Herrman behind the brand. Trafalgar LED watches were popular in the UK from around 1973 due to their relatively low cost and variety of stylish cases. To keep costs down, all the watch components were bought in and then assembled

Popular LED watch from UK brand Trafalgar.

A more futuristic-looking Trafalgar model.

in the UK. The display and chip were sourced from California and then assembled into modules in Bangkok, whilst the cases were Swiss made. Sales peaked in 1975 when imports of digital watches into the UK could not keep up with demand. Many were sold through supermarkets such as Tesco.

However, the demand for LED watches was short-lived and the technology moved on. Willy Herrman thought that battery technology would improve and allow the watches to display the time continuously, but this didn't happen. Instead, LCD displays took over and Trafalgar started selling LCD watches from 1977.

Benrus (1975)

The Benrus Watch Company was founded in New York City in 1921 by three brothers, Oscar, Benjamin and Ralph, and the name Benrus was a combination of Benjamin Lazrus' first and last names. The Second World War years were a period of gigantic growth for Benrus and, like many other watch companies of the time, they manufactured watches for servicemen. They also had top-secret government contracts to manufacture timing devices for use in bombs and weapons. The civilian watch market was also lucrative due to the limited supply of affordable watches, and consumers snapped up whatever watches Benrus could make. By the beginning of the 1950s, Benrus had become the third largest watch company in the United States, behind Bulova and Elgin, and had overtaken Hamilton. Their success came from selling mid-priced watches and only using Swiss movements.

The company was sold in 1967 to Victor Kiam of Remington Razors, made famous in his advertisement 'I liked the shaver so much, I bought the company'. Unfortunately, this was the era of the cheap Japanese quartz movement and inexpensive Timex watches, which hurt most US producers of mechanical watches. Under new ownership, the company was stripped of its most valuable assets, which were sold off and Benrus was used primarily for its strong brand name recognition. They sold a range of digital LED watch styles in the UK from around 1975, using imported components. One of their outlets was the store Dixons, which sold household electrical and electronic goods at the time. However, if your watch went wrong, the paperwork that came with your UK watch told owners to send them all the way to their service centre in Ridgefield, Connecticut!

The watches typically had two buttons: one for hours and minutes, then seconds if it was held pressed in. The other button would display month and date together. By the mid-1970s Benrus Incorporated was a diversified manufacturer of a number of consumer products,

A watch from US company Benrus.

A cheaper Benrus model.

including Destino Ltd, which sold Christian Dior Products. An attempt to consolidate all the various manufacturing enterprises under one roof proved to be a much more expensive move than anyone calculated, and a final blow to the company, which subsequently filed for bankruptcy in 1977.

Commodore Time Master (1976)

Commodore International, or Commodore Business Machines (CBM) as they were known in the UK, are probably best known for their highly successful range of personal computers in the 1980s. Founded in Canada in 1958 by Jack Tramiel and Manfred Kapp, their company started off manufacturing typewriters and expanded rapidly. By 1965 they had moved on to adding machines, and by the late 1960s Commodore were selling Japanese calculators from the likes of Casio under their own name. They went on to manufacture their own successful range of calculators, and when the global boom meant that chips were hard to

The Time Master from Commodore. (Courtesy Michael Bolton)

come by, Commodore purchased the chip manufacturer MOS in 1973 to guarantee supply. As well as electronic calculators, Commodore produced many other electronic products throughout the 1970s such as the chess computer ChessMate in 1976 and, of course, they made digital watches. The first watches were launched in 1975 using Swiss-made cases and assembled in the US with US electronics. From 1976, Commodore sold the Time Master, which was assembled in France. To keep prices competitive, Commodore used cheaper, smaller LEDs with a magnifying lens over each digit to compensate.

In the mid-1970s, Commodore had the foresight to see that personal computers were going to be the next big thing in consumer electronics and owning MOS meant that they could produce their own chips relatively cheaply. They developed one of the first personal computers, the Commodore PET (Personal Electronic Transactor) which debuted in 1977. Commodore went on to make many successful home computers such as the Commodore 64, the US's best-selling computer, before they were wound up in 1994.

Sinclair Black Watch (1975)

The Black Watch from Sir Clive Sinclair's electronics company was the only UK-designed digital watch of the period. The Cambridgeshire-based company Sinclair Radionics was formed in 1961 and specialised in build-it-yourself radio and hi-fi kits for the hobbyist. The company went on to specialise in miniaturisation, and the Micromatic radio which sold from 1967 to 1971 claimed to be the 'world's smallest radio'. In 1972, Sinclair Radionics turned their hand to pocket calculators with the Sinclair Executive. This innovative model met their trademark claim of 'the smallest in the world' and won awards for its slim design. A year later, in 1973, Sinclair launched the Cambridge range of budget pocket-sized devices which were available pre-built or as a kit.

When the Black Watch was launched in 1975, it was a somewhat risky business decision, as the digital watch market was transitioning to LCD with their much longer battery lives. However, the Black Watch design was unlike anything else on the market. As the name implied, it was all-black in colour and the name was also a pun on the Black Watch Scottish regiment. Without any visible buttons, it had a rather cool though chunky design. Made entirely from black moulded plastic, it could display either the hours and minutes or the seconds. There was also a less-common grey version which additionally displayed the date. The area beneath the display concealed two buttons operated by pressing on the left or right

The Black Watch by Sinclair.

Build-it-yourself was an option.

side. The left gave standard brightness whilst the right was enhanced brightness for use in sunlight. Pressing a button once displayed the hours and minutes. A second press showed the date, and a third press displayed seconds. There was also a setting button on the back.

Although it was an innovative design, the Black Watch was fraught with problems and a commercial disaster for Sinclair. Sinclair liked to make his products available as sell-assembly as well as ready-built, and in this case the saving was £9. However, the kits were almost impossible for the hobbyist to build – they had problems enough in the factory where they were made. The case was too difficult to clip together, the electronics were unstable and the control boards would often fail. The ready-built watches were no better. The accuracy was very sensitive to temperature changes, so the watch ran at different rates depending on whether it was being worn or not. Worst of all, the battery life was very poor and reputedly only around ten days. Many thousands stopped working and were returned at great cost to Sinclair, which made a loss of £355,000 in 1975–76 on a £5 million turnover. It was a major blow to Sinclair's finances, but fortunately the National Economic Board came to Sinclair's rescue and bailed him out. Sir Clive was a big supporter of Margaret Thatcher, and she eventually knighted him, but with her dislike of state aid she abolished the NEB when she came to power in 1979. Without the bail out we might never have had Sinclair's major contributions to the world of home computers with the likes of the ZX80 and Sinclair Spectrum, which would have been a great shame. With very few working Black Watches surviving today, if you do see a working one for sale it would be a good investment if the price is less than a couple of hundred pounds.

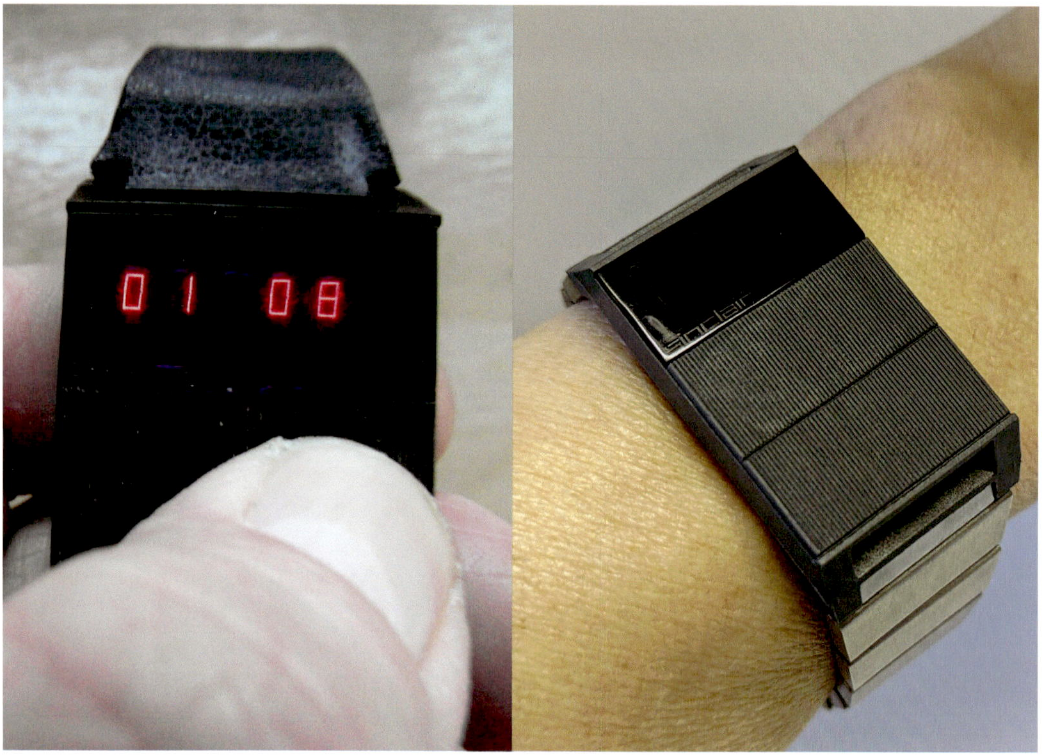

Available in black or grey.

Macy's (1975)

Macy's is a luxury US department store chain founded in 1858 and headquartered in New York. It is a household name in the US with 510 stores; the flagship store in Manhattan takes up a whole block and is the world's largest. This mid-1970s watch sold by Macy's was unusual in that it has touch-sensitive areas set into the watch face instead of the normal buttons on the sides.

Their jewellery department had high-end watches made for them in East Asia, and this particular watch would have been quite expensive at the time with its silver plating and silver mesh bracelet. The case was 35 mm in diameter and the absence of buttons allowed for a slim 5-mm profile. Touching the larger contact on the face showed the time, date and seconds on successive touches. The smaller area was used in setting the time. When one of the metal areas was touched, a circuit was made through the wearer's body via the back of the watch and through the wrist. A current then flowed which resulted in display of the time. Fortunately, due to the tiny amount of current involved, there was no electric shock! However, for it to work, there had to be good electrical contact and so a bit of sweat helped as the salt provided extra conductivity. They were always a bit temperamental, and so touch-control watches were uncommon. However, touch-sensitive technology was eventually perfected, and is now used in modern smartwatches.

Watch by Macy's without conventional buttons.

Timex SSQ 200/225 (1977)

The Timex brand was conceived in 1950 as a cheaper alternative to higher-priced Swiss watches. The United States Time Corporation had a history of making precision-made timing fuses during the Second World War, and its president Joakim Lehmkuhl had the idea of using similar manufacturing techniques to make reliable watches to rival Swiss. Watches at the time used expensive jewels for the watch's bearings and these were replaced by a hard, cheaper alloy called Armalloy. Success of the Timex brand led to United States Time Corporation being renamed Timex Corporation in 1969.

In the 1970s, with the arrival of digital quartz watches from East Asia, America's mechanical watch and clock industries were devastated. Timex closed most of its factories and reduced its workforce from 30,000 to just 6,000. Like most watchmakers, Timex sub-contracted manufacture of the case and electronics and had them assembled in Taiwan with the Timex branding. In 1977 Timex marketed a range of LED watches for men and women called SSQ (Solid State Quartz). Watch face options were round, square or oval, in either silver or gold finishes. Advertising called them 'Time On Command', which just meant that you had to press a button to see the time and distinguished them from their range of always-on LCD watches. This model was advertised at $32.95 at the time, which is around ten times that amount in today's money. As this model had an alphanumeric display, it could also display the day of the week.

Above: A Timex watch from their SSQ range.

Left: Days of the week could be displayed.

Today Timex are owned by a Dutch holding company called Timex Group, which owns many watch brands, including luxury Italian brand Vertime and high-end brand Guess/Gc. Timex also make GPS and exercise watches.

Hamilton PSR (2020)

Although LED watches were effectively obsolete by the mid- to late 1970s, they still live on today. In 2020, to celebrate fifty years since the first Pulsar, Hamilton, now owned by Swatch (formerly SSIH), launched a modern version of the original Pulsar P2, called the Hamilton

PSR. Although the case retains the original chunky cushion shape and the display is bright LED, the module has been redesigned using both LCD and emissive OLED (Organic Light Emitting Diode) technologies. In normal daytime use, the time is displayed via red LCD and is visible even in direct sunlight. But when the ambient light grows dim, a press of the button on the side changes the display to traditional red LED reminiscent of its fifty-year-old cousin. This gives the PSR as long a battery life as an LCD watch. As well as a stainless-steel version and limited edition in gold plate, there are also other models such as the special edition *The Matrix* version in all-black with green LED, selling for premium prices.

Above: Modern Hamilton PSR with red LCD display.

Right: A press of the button displays in retro LED.

2

Liquid-Crystal Display (LCD) Watches

By 1977 – only five years after the first digital watch had gone on sale – LED watches had about had their day. Their demise happened for two reasons: firstly, a button had to be pressed to display the time, and secondly, their battery life was rather poor. A revolution similar to that of the first digital watches came about with the invention of the LCD display. The first practical type of Liquid Crystal Display was called the Twisted Nematic Field Effect

A basic LCD watch by Electron.

(TN) in around 1972. There was an earlier type called Dynamic Scattering Mode, but it needed a backlight to be visible and was therefore impractical. A few watches with TN displays were marketed from 1972 onwards, but these were prohibitively expensive. However, from 1975, LCD watches became widely available and made their LED cousins obsolete. Not only was the time display always visible, the batteries lasted many times longer than LED watches. Also, smaller batteries meant slimmer-profile watches were possible.

Watches soon started to appear with extra features beyond just the time. The date and the day of the week came first. Soon after a stopwatch, or chronograph, was added. Then many other functions such as alarms, timers, dual time and world time zones. This section looks at some of the milestones in LCD watch design.

Seiko 'O' Series (1973)

The Japanese company Seiko were a prolific manufacturer of LCD watches in the 1970s and 1980s. The company was founded in Tokyo in 1881 by Kintarō Hattori and, in the century and a half since, it has grown into a multi-billion-yen corporation. Today, as well as clocks and watches, Seiko manufactures, amongst other things, musical instruments, semiconductors, printers and jewellery. But in the 1950s and 1960s, Seiko produced a number of world firsts in watch design.

The year 1956 saw the first self-winding watch, the Automatic, but at three times the price of a wind-up watch the market for them was limited. Three years later, Seiko invented a new self-winding mechanism called the Magic Lever. This revolutionary device used a lever and claw arrangement to harness the energy from an internal mass in a way that was both simple and efficient. The new mechanism instantly popularised self-winding watches and is still used by Seiko today. Seiko's trademark was always precision time keeping and to celebrate this, in 1960 they created the Grand Seiko, the most accurate mechanical wristwatch in the world.

In the early 1960s Seiko experimented with quartz clocks. Though early models were wardrobe-sized machines, the Tokyo Olympics in 1964 was an opportunity to showcase a dramatically smaller quartz clock which was portable and ran on batteries. Seiko also provided all the stopwatches for the games.

Perhaps Seiko's most significant achievement though was developing the world's first quartz watch. Launched in Tokyo on Christmas day 1969, the Astron had a claimed accuracy of five seconds per month – five times better than the best mechanical watch. It would run for a year before needing a battery change, but the downside was that it cost as much as a medium-sized car. Also, it still had conventional hour, minute and second hands.

It was 1973 before Seiko brought out their first all-digital watch. Seiko were unusual in that they bypassed LED watches completely – they went straight from quartz-powered mechanical to LCD digital. Seiko weren't the first LCD watchmaker, but they were the first to produce a watch that displayed digital seconds. This was the 0614 model, the first in a line of watches designated the 'O' series. It attracted worldwide attention when it was marketed in October 1973. The early LCD display had a yellow filter to protect against UV and had a backlight for night-time. Note that the segment arrangement making up the digits is of a slightly different design to what came later and shows particularly with the '4'.

Seiko's first LCD, the 0614.

Seiko 0664 with side crown wheel.

Another watch from the 'O' series, the 0664, was produced in 1975. It was an attractive watch with its black face and gold framing of the time window, and considerably more affordable than the 0614. By this time, LCD displays were more resilient to UV light and so didn't need the yellow filter anymore. The 0664 just displayed the time and had a crown wheel on the side just like a mechanical watch. Pulling out the crown put the watch in setting mode and the seconds flashed. The right-hand button then selected which of hours, minutes or seconds to adjust. The left-hand button then advanced the hours or minutes, or set the seconds to zero. In time mode the left-hand button operated the backlight.

Also in 1975, the 0634 was released and was the world's first chronograph digital watch. It came in three different configurations – this one being the 0634-5009 – and all were heavy with thick profiles of around 12 mm. Like the 0664, it had a crown on the right-hand side which when pulled out, the time could be set with the front buttons. When pushed in, it operated as a button to switch between time and stopwatch modes. The front right-hand button started and stopped the stopwatch and the front left button either reset the lap or measured a lap time. In time mode, the front left button operated the backlight.

Seiko 0634-5009 in stopwatch mode.

Seiko 'A' Series

From 1976 Seiko manufactured a whole range of different models with alarms, known as the 'A' series. This model, the A159, dates from around 1979 and has both alarm and chronograph functions. Many of the 'A' series have the days of the week permanently displayed via letters on the glass, and a bar over the top indicates the current day. The labels around the watch face indicate what the buttons do in its various modes, which made it easy to operate. For example, to set the alarm, the mode button (bottom left) was pressed twice to put the watch in Alarm Set mode, as indicated by a flashing bar above 'Mo'. Then the Select button (top right) and Set button (bottom right) were used to set the alarm time and either AM or PM. Alarm On was indicated by the cone icon in the top right of the display. In stopwatch mode the time was displayed with tenths-of-a-second accuracy, and the lap function froze the time and was indicated by a flashing bar over 'Th'.

Seiko model A159-4029 with alarm.

The A159 came in several variants, this one having a two-tone gold/steel bracelet, but others had a squarer case style and a steel/black bracelet. The A159 required a single battery accessed by a hatch in the case back.

This later model – the A174 – has similar functionality to the A159, but a more complex display with five separate LCD areas. The top area displays the time, day and date. Below that is the alarm time with a speaker icon to the left showing that the alarm is set. Below the alarm area is the stopwatch time showing minutes, seconds and one-hundredths of a second. The cool thing about this model is the figure to the left of the stopwatch who runs whilst it is counting. Seiko went to the trouble of animating three different arm and leg positions to give the runner a more realistic motion. Having these separate areas meant that everything was visible at the same time without having to press any buttons. The watch still looks quite cool forty years on, especially with the unusual gold tint to the LCD background.

A714-5050 'running man' model.

Seiko 'G' Series

In the 1980s, LCD displays became more complex as manufacturers sought to make their watches stand out from the crowd. Seiko produced this popular and striking-looking watch, known as the Digiborg in Japan but formally designated the G757. The cases came in two different shapes: the one here has the horseshoe face and is the most collectable because the G757-5020 Sports 100 model appeared on 007's wrist in the 1981 Bond film *Octopussy*. It appears many times throughout the film, for example in Q's lab where Q explains how Bond's watch can track the bug in the Fabergé Egg: 'The homing device is compatible with the standard issue radio directional finder in your watch ... if you haven't lost it.' So it appears the G757 is a standard issue watch for secret agents, and this explains why Bond was already wearing it throughout the first scenes. In the film, the watch is shown with a red light moving around the seconds area, which the real G757 obviously doesn't have. Disappointingly, it doesn't have the radio tracker either!

The one in the photo differs from Bond's only in that it has the words 'Silver Wave' on the top of the face instead of 'Sports 100' and doesn't have a metal bracelet. Silver Wave was the name used by Seiko for their water-resistant models and each had the Tsunami logo on the back. It featured an hourly chime, dual time (where one time could be viewed in analogue), an alarm, a stopwatch and a countdown timer. The large display has a

Seiko's G757 featured in James Bond's *Octopussy*.

Seiko G757-405A with analogue display area.

stylised clock face which shows the seconds graphically, and to the right indicates which of the five modes is selected.

This G757 model, the 405A, has the other square-style case, but has very similar functionality to the 5000 model. These watches go for high prices, not just for the Bond connection but because they are arguably the best digital models that Seiko produced. Sadly, they are prone to failure if the battery is left in and goes flat, but if you can find a working one, it would cost you several hundred pounds, and an actual Sports 100 model upwards of £1,000.

Seiko 'D' Series

In 1980, Seiko created their first dot matrix LCD display in a watch called the D031. Dot matrix displays are made up of individual pixels and so not only can words be displayed, but graphics effects as well. The D031 had a basic display but when a side button was pressed, the date slid off the screen and the time slid on, plus it looked very cool! This was the start of their 'D' series of increasingly complex watch displays that dot matrix allowed for. In 1981 Seiko released this D229 model, also a Silver Wave version. The four buttons on the front selected the mode required – alarm, timer, stopwatch or time and date display. The top half of the display was the dot matrix area where the name of the mode was displayed or, in time display, the day of the week. The lower half showed the digits for hours and minutes in time mode or minutes and seconds when the stopwatch

Above: D229-5000 showing time and stopwatch modes.

Left: Seiko D138 'running man' watch.

or countdown timer were in use. There were also three separate alarms, the alarm being indicated by a bell on the screen. The right-hand control was a crown wheel which could be pulled out and used to set the time or turn the alarms on or off.

The D138 is perhaps Seiko's most iconic digital watch, better known as the 'running man' or 'big running man' to differentiate it from the A714. It dates from 1982 and has an almost identical case to the G757, but Seiko used this watch to show off their dot matrix technology through animation. In stopwatch mode, the dot matrix area in the bottom right displays an animation of a man sprinting. It can also display up to five different pictures, which are used as reminders for the two alarms that can be set. For instance, the boy-and-girl icon is for a reminder of a personal meeting. The other icons are a bell, a telephone, a plane and a man at a desk, which should all be self-explanatory. Whilst showing the time, the dot

Six different icons were available for use as reminders.

matrix area displays the day of the week and the date. Like the G757, these are also very sought after and, because the dot matrix displays don't survive intact very often, they are very rare to find in a working condition.

Seiko S760 (1995)

This watch was part of Seiko's 'S' series, standing for solar-powered, but its other feature was that it contained a radio receiver. Before smartwatches came along, which can synchronise the time with your phone via Bluetooth, the only way for a watch to auto-set the time was via a radio signal. There are a handful of transmitters around the world which broadcast time signals, and this watch could pick up signals from the England transmitter if you lived in the UK. Having a radio receiver inside it made it a heavy, chunky watch with a rather wide display, but also easy to read. The lower part of the display is dot matrix to allow the display of letters for all the world city abbreviations. The display is reversed, i.e. white on black, and the watch face also doubles as a solar cell to keep the battery going for a claimed eleven months between charges. Other features are a stopwatch, dual time display, world time, three alarms, automatic calendar and water resistance to 100 m. Overall the watch has a feeling of high quality, extending to the stainless-steel bracelet which has a push-button clasp release.

Seiko S760-0AB0 showing the time in Rio de Janeiro.

Citizen Ana-Digi (1979)

The Citizen Watch Co. has been around since 1930, originally founded in Switzerland, but today forms part of a global group of companies based in Tokyo. It is one of the world's largest producers of watches. Citizen were behind many innovations in time keeping throughout the twentieth century, such as Japan's first fully waterproof wristwatch in 1959 called the Parawater. This led to a range of diver's watches culminating in the world's most pressure-resistant watch in 1982 which could stand a depth of 1,300 m. In 1975 Citizen created the most accurate watch in the world, the Crystron Mega, and a year later, the first solar analogue watch. The year 1978 saw the record for the slimmest watch set by Citizen at under 1 mm, and 1979, the smallest watch movement.

In the 1970s Citizen produced the Crystron range, with many different models of increasing complexity as the decade progressed. This chronograph model from 1979 is an early example of what is known as an analogue-digital or ana-digi watch which squeezes two LCD displays and an analogue clock into the watch face. The top LCD display shows either the current time or minutes accumulated when in stopwatch mode. Below that is a two-digit LCD which displays either the date or one-hundredths of a second for the stopwatch. To the left is the analogue clock for those who prefer to see their time displayed with hands.

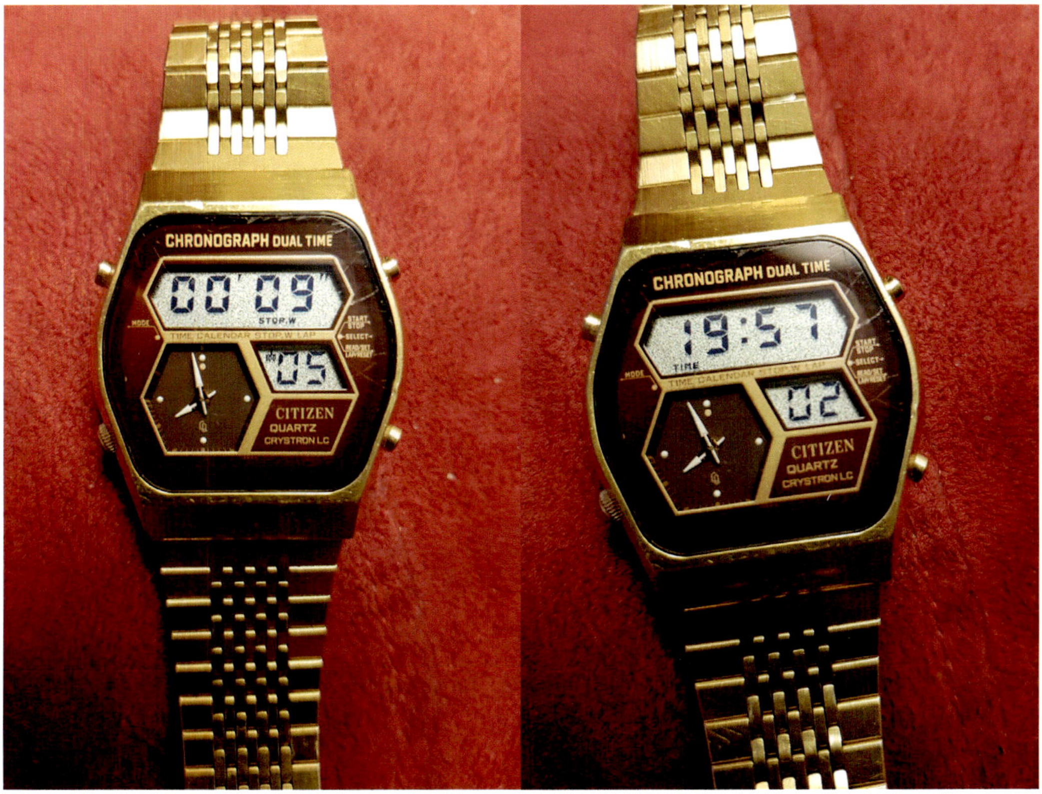

Citizen Crystron model in stopwatch and time modes.

Citizen advert showing models from their Crystron range.

Citizen Ana-Digi Temp (1980)

In 1980, Citizen took the Ana-Digi concept and supercharged it, packing as many features as they could think of into this show-off piece of tech. It has been referred to as the Apple Watch of its day, costing about a million dollars to develop at the time. The totally new feature was that of measuring temperature in either Celsius or Fahrenheit – a first for a watch, although the watch needed to be removed from the wrist for an accurate measurement. The two analogue clocks were multi-function, with the left displaying hours and minutes and the right either seconds, alarm, stopwatch or dual time. In dual time, the left would display GMT and the right would indicate the hours in another time zone. An arrow in the bottom left display pointed to the watch's current mode. The middle LCD displayed either time, alarm, dual time, stopwatch minutes and seconds or day and date. The calendar was fully automatic, dealing with leap years up to 2019. By then Citizen figured you would have bought one of their newer models! The lower LCD displays temperature or one-thousandths of a second depending on the mode. There were several styles and colours available apart from gold, and perhaps the best looking was in black with silver dials.

Citizen Ana-Digi Temp watch.

A slightly later Ana-Digi Temp model.

Nowadays, due to the popularity of all things retro from the 1980s, Citizen make new Ana-Digi Temp watches, exactly faithful to the original looks, and calling them 'Vintage Retro'. Good design never goes out of fashion.

Pulsar Y486 (1981)

The Pulsar brand had been used very successfully in the early 1970s to sell the first LED digital watches, before it was sold to Seiko in 1978, who keep it to this day. Back in 1981, Pulsar offered this LCD watch, the Y486-4030 with alarm and chronograph features. The display is pretty basic, at least compared with the Citizen models of the time, and comprises six digits for both time and stopwatch display, and a row of indicators along the bottom of the screen labelled

Early Pulsar Alarm Chronograph model.

zero to nine. Zero to six are used to indicate the current day of the week, seven indicates the alarm is set, eight for PM and nine indicates stopwatch mode. When the stopwatch is running, the indicators sweep along the bottom, right to left in a simple animation.

Pulsar Y789 (1982)

A year later in 1982, Pulsar released this Y789-5059 model with similar functionality to the Y486 and with a particularly clear, easy-to-read face. Function-wise it had an hourly chime and wake-up alarm, stopwatch with lap timer (each of which has an icon to indicate operation), date, day, twelve- or twenty-four-hour display and a backlight.

A later model using graphics for the watch modes.

One really nice feature fixes the problem that metal bracelets have with sizing. Normally links have to be added or removed, but on this watch the clasp can be moved to the appropriate position on the bracelet and then locked in place.

Casio AL-180 (1987)

No book on digital watches would be complete without mentioning the name Casio. Today the Casio Computer Co. is a multi-billion dollar electronics giant based in Tokyo, and perhaps best known for its calculators, electronic keyboards and affordable digital watches.

In the 1980s, Casio sold a huge variety of digital watches, owning the budget end of the market. Many were made primarily out of resin, thus enabling easier mass production. Today, the watchmaker has hundreds of different models on the market, but it all started back in 1974 with the Casiotron. Although Seiko had released their first LCD watch a year earlier in 1973, Casio's offering had something extra. Not only did the Casiotron tell the time on an LCD screen but was the first digital watch to determine the days in each month, known as an automatic calendar. The various models in the Casiotron range were so popular at the time that Casio decided to re-release them in the 1990s.

A Casio solar-powered model.

Most watch batteries need replacing after a year or so, but Casio attempted to eliminate the battery altogether in the AL-180 model from 1987. Inside, there is what looks like a normal coin-sized battery but in fact it is a type of capacitor. In conjunction with the solar cell on the front, the watch can be charged with four hours of sunlight. There is a sun symbol on the display which disappears when fully charged. The watch has a twelve- or twenty-four-hour display option, date and day of the week and an alarm. There is no back light in order to conserve power. As with the Casiotron, its popularity caused it to be re-released in the late 2000s, as well as being pirated by other companies. In yet another retro-revival, Casio released an updated version, the AL-190 in 2011, very similar in style to the AL-180, but with additional features such as a stopwatch, and choice of colours in a resin case.

Casio A500 (2014)

Watches which could tell you the time anywhere in the world were popular for a time in the 1980s. Casio offered the A300 model in 1986, which had a world map on the display and a line going from north to south would indicate which time zone was currently selected. This was another watch that Casio thought was so good at the time, they would re-imagine it with a modern version known as the A500, available since 2014.

Casio A500 world time model.

The A500 looks very similar to the original, but has many more features inside. Pressing the Search button steps through forty-eight different cities in thirty-one time zones. There is a one-hundredth-second stopwatch with twenty-four-hour measuring, split-time and first and second place times. There is also a multi-function countdown timer, five alarms, full auto calendar until 2099 and even an LED backlight with selectable illumination duration!

Casio F-91W (1989)

The F-91W, released in 1989, was highly popular for its low price and long battery life. In 2011, it claimed the record for the best-selling watch when production hit more than three million per year. The F-91W also has a darker claim to fame. It gained a reputation for being the watch of choice for making timers in terrorist bomb plots.

Casio's hugely popular F91-W.

The Casio Illuminator with back-lit display.

Casio Illuminator A200 Futurist (1997)

This model, the A200, was part of Casio's Futurist range and its USP was its light-up display which gave the model the alternative name of Illuminator. Most LCD watches had a light at the left-hand side which didn't do a particularly good job of showing the time in the dark. The Illuminator though lit from behind, giving an even glow in a pleasant blueish-green colour. The rather large LIGHT button on the front also made it easy to find and press in the dark. Apart from the light feature, the watch had a daily alarm, hourly chime and stopwatch. This model had a reverse display, i.e. white-on-black instead of the more conventional black-on-white, known as a negative display. Even though the buttons and case are plastic, which gave it a cheap feel, it was an easy-to-use and popular watch.

Casio Poptone LDF-50 (2011)

Casio introduced this colourful model in 2011 which was aimed at younger, more fashion-minded customers of both sexes. The display and four chunky buttons on the front are designed to look like they have been inserted into the silicone wristband, which contrasts with the traditional square, metal-cased models. The display is also slightly curved, with dimensions of 23 mm by 40 mm. Other more subtle colours were also available for less flamboyant buyers. Although it looks like it should be a fitness tracker, it is actually a fairly standard chronograph watch, with good water resistance up to 5 bar. Its large, easy-to-read

The brightly coloured Casio Poptone.

display shows hours, minutes, day, date and month all at a glance. Below these is a graphical area where circles are used to illustrate a continuous ten-second countdown timer. Other features are an alarm, hourly chime, stopwatch with one-hundredth-second precision and an electro-luminescent backlight with variable time delay. A CR1216 battery gave up to three years of light use. Maybe not to everyone's taste but, costing only £20 at launch, it was good value and filled a niche in Casio's portfolio of affordable watches for everyone.

Casio G-Shock (2016)

For those who need their watch to be almost indestructible, Casio provides the G-Shock range, standing for 'Gravitational Shock'. These are knock-resistant watches aimed at those who enjoy outdoor activities and sports, and also the military. The concept goes back to 1981, and since then, many models have been added to the range and feature waterproofness, shock resistance, an alarm and either digital or analogue display.

This 2016 model, the GW-7900B-1ER, has an inverse display and a large case of 52.4-mm diameter by 50-mm lug-to-lug by 17.7-mm deep. Just by looking at the rugged styling, you can believe Casio's claim that it can be dropped from a 10-m height without damage. The case is made of a urethane resin, so it's not brittle like plastic and, as well as shock, the watches are tested for water, ice and vibration resistance.

The 'G' button provides a bright green illumination when pressed. It can also be set to light up on a wrist movement. The middle of the display is the unusual part and has three separate graphic areas. The left side shows the phases of the moon in thirty-two different graphics to cover the twenty-eight-day lunar cycle. In the centre is tidal information – the height of the wave graphic indicates whether the tide is spring (big tidal difference), intermediate, or neap (small tidal difference). There are six segments in the peak and trough to indicate whether the local tide is low or high. The circular graphic in the middle right indicates seconds in both time and countdown modes. The whole of the display is a solar cell and the battery lasts nine months from a full charge without any sunlight. One of the most interesting features is the ability to synchronise time to a broadcast radio signal. Most of Europe is covered by just two transmitters – Anthorn in the UK and another in Germany. The manual suggests that receiving a signal may be hit and miss if you live in a built-up area and, to synchronise, place the watch on a window sill at 45 degrees and at night! As well as auto-time setup, the watch has five alarms, a stopwatch, countdown timer and world time with forty-nine cities to choose from. In 2017, Casio passed the 100-millionth G-Shock sale.

Casio G-Shock GW-7900B-1ER.

3

More Than Just a Watch

As we have seen, the technology inside the electronic watch never stands still for long. Almost as soon as the basic watch was developed, a calendar was added. Not long after, a stopwatch, then an alarm. Electronics is a constantly changing technology where every year more and more transistors can be squeezed into the same space. So, every time the manufacturers produce a new model, they have more capabilities, which keep the

Casio CA-53 calculator watch from 1988.

customers buying. In the 1980s and 1990s, watch manufacturers offered features such as radio, spoken time, games and even TV for our wrists. Of course, most of these were produced for their novelty value rather than serious applications and never sold in big numbers. But they were a showcase for the technology of the time.

This chapter looks at some of the more unusual things that you could do with your digital watch in addition to telling the time.

Calculator Watches

Pulsar Time Computer Calculator (1975)

Pulsar, the brand name behind the Time Computer Company, was the first to cram all of the electronics from a calculator into a watch-sized case. The first models were so difficult and expensive to make that, like the first Pulsar watch, they were made from 18K gold and aimed at the most affluent customers. A limited edition of 100 pieces went on sale in upmarket New York stores in time for Christmas 1975, priced at an eye-watering $3,950 (approximately £24,000 in today's money). It was said at the time that President Ford wanted one but his wife said 'No way'!

However, by the spring of 1976, Pulsar released a more reasonably priced stainless-steel version costing $550. Included with the purchase was a bespoke Pulsar-branded pen, one end of which served as a stylus to operate the buttons. These first models contained what was known as the '901' module and, like Pulsar's previous digital watches, they were a high-quality product.

Pulsar Time Computer Calculator in solid 18K gold. (Courtesy Lloyd Harvey)

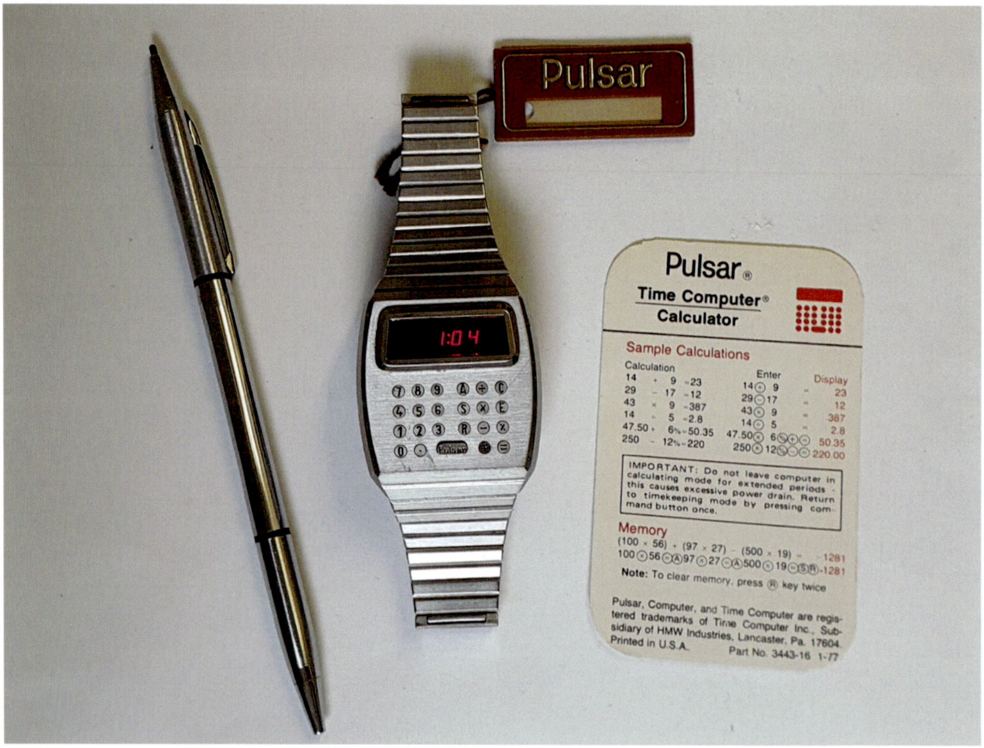

Useful pen-stylus for the tiny buttons.

Pressing the command button (labelled Pulsar) displayed the time, which had an accuracy of sixty seconds a year. A second press displayed the month and date with automatic month adjustment. To operate the calculator, first the '0' was pressed (using the stylus as the buttons were far too small for fingers), which then showed a zero on the display and you were ready to go. The calculator had a six-digit display but calculated to twelve digits. As well as the arithmetic functions, there was per cent, memory add and memory subtract, and floating-point decimals.

The watch required four batteries to give it a reasonable battery life (although it would run on two), and these contributed to the watch's not inconsiderable weight.

In 1977 a newer, improved '902' electronics module was introduced which gradually replaced the older ones. These had a slightly different keyboard (the 'R' and 'S' and the 'C' and '=' keys were transposed), a new leap year calendar and a flick wrist feature to display the time without having to press a button. The case options remained the same – either stainless steel, 14K gold plated or 18K solid gold.

Other features of the watch that the owner's manual mentioned were water-resistance, shock-resistance, anti-magnetism and maintenance-free. To quote the manual on maintenance: 'This is now obsolete, the "movement" now being replaced by a "brain". As a result there is no wear, no oil, no servicing and no repair costs'! This was true in one sense, however: the electronics were delicate and complex, and so they would fail now and again, even though Time Computer Co. supplied a three-year guarantee.

The six-digit calculator display.

The Pulsar calculator watch never sold in big numbers, being both expensive and a niche market, and so today they are very desirable amongst watch collectors. Working watches with their boxes can easily fetch over £1,000.

Other manufacturers were quick to release their own calculator models. Hewlett-Packard produced the HP-01 and Compu Chron marketed the Hughes Aircraft Calculator. Sir Clive Sinclair also produced a small number of just wrist calculators (no time functions). But LED calculator watches were on sale for only a short time before cheaper, more efficient LCD calculator watches came along to make them obsolete.

Seiko 'C' Series

Just a couple of years after the first LED calculator watch, Seiko launched the first in their 'C' (Calculator) series of LCD watches. The C153 combined basic watch functions with a calculator keypad of six by four raised and dimpled buttons beneath the display. On the side of the watch, the middle button of the three was a crown type, which pulled out to set the time. The lower button worked the light and the upper button displayed the date and seconds. At the rear, a circular hatch with a coin slot gave access to the single battery.

The watch was quite large at 42 mm by 35 mm and 11 mm deep, but the wide face allowed for the eight-digit calculator display. Function-wise, there was a memory, square root and per cent as well as a sign change button. The tiny buttons had to be pushed with a sharp implement, namely a ballpoint pen, and the recesses on each button were there

Left: Seiko's C153-5011 calculator watch.

Below: The Seiko C359 calculator with even smaller buttons.

to stop the pen slipping. In its three or so years of production, there were a few different models produced in the C153 range, including watches with faces in black or gold.

Seiko released their second 'C' calculator watch, the C359, two years later in May 1979. There were a few differences with the previous C153 model, most notably it being smaller both in face size and depth. There were now two rows of calculator buttons below the display instead of four, but this still provided your five basic calculator functions, which was what most people wanted. The per cent and clear functions hadn't been removed, just moved to the buttons on the right side of the watch. The display was eight digits plus floating point and an 'E' indicator for overflow. Watch functions included an alarm, hourly time signal and days of the week. You could also light all the segments in the LCD to test them by advancing to set mode then pressing both right-hand buttons.

The big drawback with the watch though was that the buttons were far too small to be used with your fingers. A ballpoint pen and a steady hand were essential for working the calculator. Even so, the pen would often slip off the buttons and many watches that have survived have the lettering scratched off by pen damage. It was also common to over-press the buttons and push them in permanently. That said, this still looks like a classy watch today with its neat buttons and stainless-steel case. They were very popular in their day and there were a few different colour variations sold, the black keyboard version being particularly common.

Pulsar Y739 (1979)

This Pulsar-branded watch was also produced in 1979 and, being owned by Seiko, shares some similarities with their models. The Y739 though is a much nicer design with its two-tone face and strap, and its almost octagonal face. Other colour variations were available such as a black keypad, but perhaps the gold/silver combination looks best. The Y739 has no

Pulsar's calculator watch.

side buttons which gives it a sleeker look, but this does however make the keypad look rather cluttered as everything, including setting the time, has to be done from there.

To help with the fiddly sized buttons, a pointed stylus was provided which fitted into the bracelet when not in use. Although it clipped in place it was not impossible for it to fall out and many for sale today are missing this key piece. The calculator functions were fairly standard, with an eight-digit display, square root, per cent and a memory on the calculator and an alarm, twelve-hour AM/PM time and backlight on the watch. Strangely, there was no date function, which is highly unusual for a digital watch of the late 1970s.

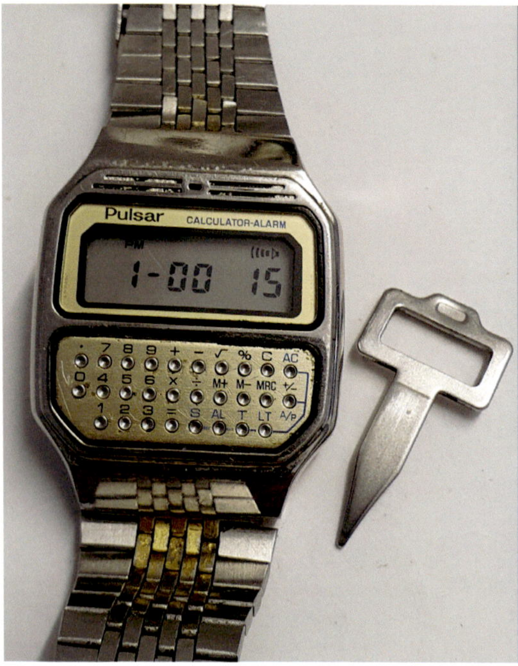

Above: Stylus fits neatly into the bracelet when not in use.

Left: Alba-branded version of the Y739.

Seiko also created the Alba brand in 1979, mainly for the Asian markets. Many of their designs were adopted from the parent company Seiko, such as this Y739 model, which is identical to the Pulsar above but with a black keyboard.

These were the last of the calculator watches that required a pen or a stylus to work them. By the early 1980s Casio released a plastic calculator watch, the CA-50, which, being a larger watch, had buttons that you could (just) operate with your fingers.

Game Watches

Casio Game-10 (1980)

The late 1970s were the beginning of the video game era. Starting off in arcades, they would spread to shops and pubs, and then the home via computers and TV consoles. Casio decided they would tap into that market via the watch. First, the company released the MG-880, a calculator with a *Space Invaders*-like mode that used numbers to depict the aliens. After that success, Casio started its range of Game watches.

The first of these was the GM-10 or GAME-10, released in 1980, only eight years after the first digital watch. The game is essentially *Space Invaders* in three lines, but played

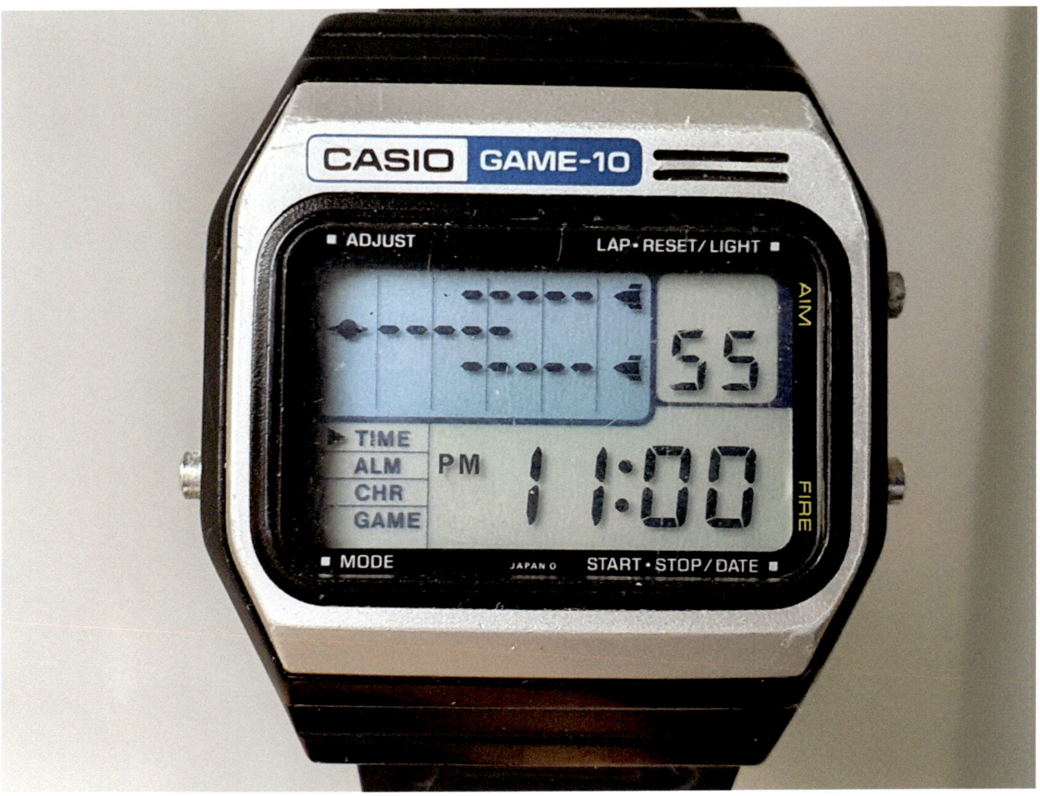

Casio's GAME-10 with built-in arcade game.

horizontally rather than vertically. In fact, the gaming aspect of this watch is quite subtle – no fancy coloured buttons or screen. Nevertheless, the game plays well and is sufficiently challenging to make players want to go back again and again to beat their high score. The idea is to move your ship (on the left of the screen) up and down using the AIM button. Once aligned with one of the aliens on the right-hand side, the FIRE button is used to stop their missiles before they destroy you. Points are scored based on where the incoming missile is when you hit it, and you get the most points for hitting the ship before it has fired. Sadly, you would not have been able to get away with playing during a school lesson due to the sound effects! As well as the game, the watch has an alarm, stopwatch, date and a light. When not in game mode, the game area plays an animation where the aliens and your ship fire at each other continuously. Casio had switched to resin case and strap by this time to make them cheaper and lighter.

The watch was popular because a second watch, the GAME-20, was released which used the same internals – the 165 module – and was essentially unchanged from the GAME-10. A third watch, the GAME-30, had a different game based on submarine warfare and a vertical shooting position. The fourth watch in the series, the GAME-40, was completely different and turned the display into a puzzle. Unsurprisingly, these are sought after by collectors for both their rarity and reasons of nostalgia and therefore prices are high.

Nelsonic *Legend Of Zelda* (1989)

The Legend of Zelda video game was an early adventure-style game by Nintendo for their NES (Nintendo Entertainment System). It was released in 1986 in Japan and 1987 elsewhere and, although not the first, it is credited with perfecting the adventure game genre. In 1989, Nintendo licensed the Nelsonic company to release this Zelda-inspired LCD game watch.

Legend of Zelda watch's tiny adventure game.

The watch has three modes – time, alarm and game. When in game mode, pressing the bottom right button starts a new game where the score is zeroed, some music plays, and you have four lives for the game. Your character is called Link and is represented by a little guy with a shield, visible in the middle of the screen. The four buttons below the screen move Link around the play area, which represents looking down into a room or dungeon and is divided into twenty-four squares. There are four rooms in each level and four levels.

In each room, Link is attacked by his enemies (iron balls, bats and dragons) and each can only be defeated by using a specific weapon (boomerang, sword and bomb, respectively) picked up one at a time. When all of Link's enemies have been destroyed in a room, a key appears allowing access to the next room. Once all the enemies in the fourth room are killed, the dungeon's boss, Dragon, will appear for a fight.

Compared to Casio's earlier Space Invaders game, Zelda was quite sophisticated. However, the tiny screen required players to have twenty-twenty vision in order to keep up with the fast gameplay, and nimble fingers were needed to manoeuvre Link around at speed to avoid the many things trying to kill him. However, at the time, a video game on a watch was super cool and 12 million game watches were sold in all.

This was one of two Zelda LCD games, the other being the confusingly titled *Zelda* (Game & Watch), which was a larger handheld game (and not a watch as well).

Promotional Watches

Texas Instruments' *Star Wars* (1977)

Blockbuster movies and franchises such as *Indiana Jones, Jurassic Park* and *Star Wars* are always accompanied by the sale of marketing and promotional items for the fans to own and collect. Traditionally, this would be a poster for your bedroom wall or an action figure,

Star Wars promotion watch from Texas Instruments.

but with the advent of the age of cheap electronic devices, fans could afford to purchase a clock, a pocket calculator or a wristwatch themed for your favourite film.

A pioneer of the ultra-cheap digital watch was the electronics giant Texas Instruments. Today, Texas Instruments Incorporated (TI) is a multi-billion-dollar technology company based in Dallas, making integrated circuits and semiconductors for global customers. They are perhaps best known outside the US as a manufacturer of electronic calculators. Formed in 1951, they initially made electronic equipment for laboratories and industry, but soon expanded into many areas of electronics, starting with the first commercial silicon transistor in 1954. In the 1960s they developed a range of transistor-transistor-logic (TTL) chips called the 7400 series which were used throughout the electronics industry for decades. In 1971 TI developed the world's first microprocessor-on-a-chip for use in its pocket calculators.

Another first was the plastic LED watch priced at under $20. Launched in 1977, it was timed to coincide with the *Star Wars* movie and was offered at cinemas during premiere screenings of the film. The watch had special packaging which included a display case and stickers of *Star Wars* characters that could be applied to the watch face. Texas Instruments also produced some other watches with special branding, e.g. for Indy car racing, as well as selling more conventional metal-cased digital watches.

Because of the film's cult science-fiction following, these watches are much sought after by fans of the saga and also for nostalgic reasons by those who once owned one. Sadly, due to their age, cheap construction and also the common problem of battery leakage, most surviving watches no longer work.

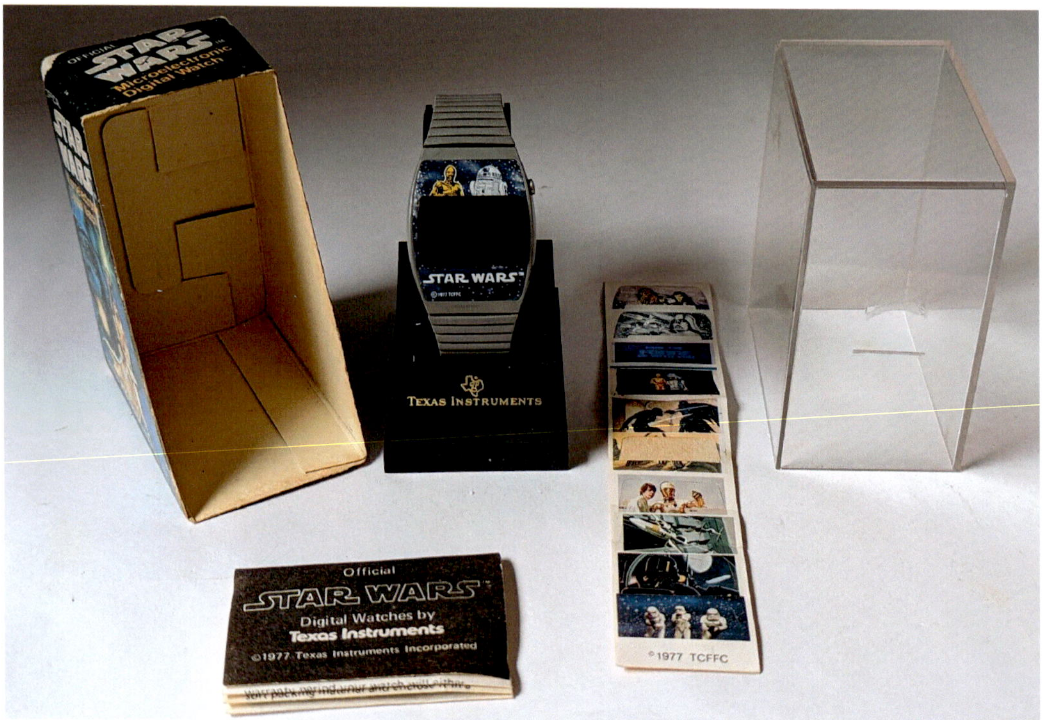

Film scenes could be applied to your watch.

Dukes of Hazzard Watch (1981)

This promotional watch from 1981 was made for fans of the hit TV series *The Dukes of Hazzard*. The show was set in the fictional Hazzard County in the US and ran from January 1979 to February 1985. It followed the various humorous escapades of cousins Bo, Luke and Daisy Duke who frequently fell foul of local sheriff Rosco P. Coltrane and corrupt commissioner Boss Hogg. One of the stars of this action-comedy show was their customised 1969 Dodge Charger nicknamed the General Lee. Merchandise for the show was provided by electronics company Unisonic Products Corporation that existed from the 1970s until the 1990s. Besides *Dukes of Hazzard*, the company made calculators and watches as well as other licensed items such as Mickey Mouse phones. In the late 1970s they made a few home video game consoles similar to Atari's Home Pong. They also made the transparent telephone that everyone seemed to own at the time. Unisonic's themed watch was pretty basic, but they really put a lot of effort into the presentation, with both box and slip case featuring plenty of images from the TV show. The series was given a second chance in the 1990s when a US TV channel revived the show for a new audience. Large quantities of unsold stock from the first time around showed up on eBay and were bought by fans and collectors. In the US, *Dukes of Hazzard* has a big following at events such as Comic Con.

Digital watches had become so cheap by this time that they were virtually disposable items and companies could make a profit selling them at $5. It was only years later that nostalgia and scarcity made them valuable.

Watch celebrating the hit TV series *The Dukes of Hazzard.*

Seiko Memo Diary (1984)

When computer fever broke out in the early 1980s, Seiko were quick to capitalise by releasing a range of watches which could store and retrieve data, such as memos, diary entries and schedules. Their first model, the D409, had a two-line display and characters could be entered by using four buttons on the side to choose characters from a list. The success of the D409 led Seiko to produce watches which came with a keyboard attachment and made text entry a whole lot easier.

Many of the models had plastic cases, but the one pictured, the UW02-0010, had a gold-finished metal case, gold-tinted display and matching gold-coloured keyboard. Text was displayed via four lines of ten characters, each composed of five by eight dots allowing both Latin and Japanese alphabets. Data was entered into the watch via the small keyboard and transmitted to the watch via induction. A coil of wire in the back of the watch and another in the keyboard allowed data to be sent by magnetic pulses at an impressive 2048 bits per second. After clipping the watch into the keyboard, the TRANSMIT button on the watch established a connection. On the keyboard, the MEMO key allowed a single page of up to 1,000 characters to be entered, or the SCH key started entry of an appointment reminder. If required, the display contrast could be adjusted by hitting either of the CNT buttons.

Seiko Memo Diary and its keyboard for data entry.

A four-line memo could be displayed.

After disconnection from the keyboard, the MODE button on the front of the watch let you check your appointments or view your memos. On entering the Memo mode, the first four lines were displayed, then the left- and right-hand buttons scrolled up and down the memo. The Schedule was viewed in much the same way – scrolling up and down appointments which were in chronological order. The MODE button was also used to set a separate alarm and display the time.

Seiko UC-3000 (1984)

This model, with its black-painted aluminium bezel, was functionally almost identical to the Memo Diary. As a watch, the large screen was useful for displaying all the time and date info on one screen – month, date, day, year, hours, minutes and seconds. The accompanying matt black keyboard, the UC-3100, required a single CR016 battery and differed from the Memo Diary version only in lacking Japanese characters.

This memo diary was the stepping stone from the digital watch to the digital personal organiser, which Psion and others were pioneering at the time. Ultimately though, tools for organising your life would be incorporated into smartphones and smartwatches.

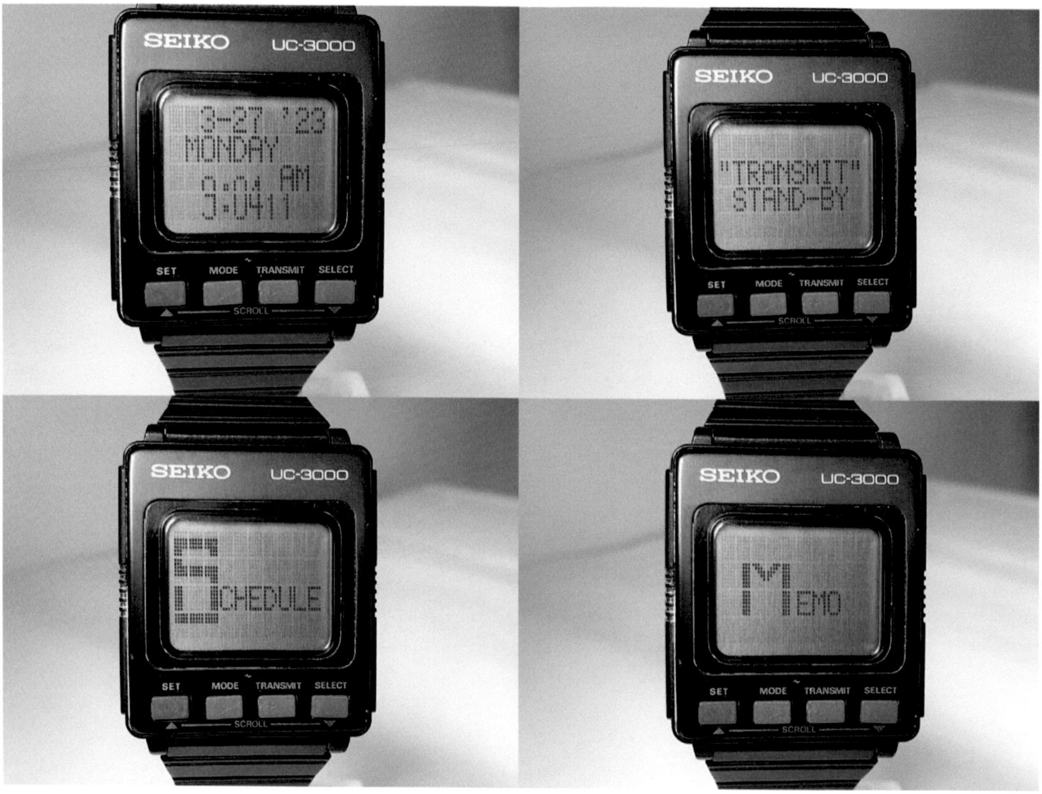

The UC-3000 model showing the various modes of operation.

Gadget Watches

Sinclair FM Radio Watch (1985)

With the Sinclair company's history of miniaturising gadgets, it was just a matter of time before they decided to try and squeeze a radio into a digital watch. Not just a simple AM receiver, but a complex FM radio! As you might imagine, trying to fit all the required components into something that could be worn on the wrist was something of a challenge. It was made doubly difficult with the watch having its own loudspeaker rather than just a socket for an earpiece.

The design they came up with split the watch into four compartments: the watch itself, the loudspeaker, the tuner and the battery compartment. The first three of these were joined together with an ingenious bellows arrangement so that the watch could be bent around the wrist. Even so, miniaturising the mechanical and electronic parts required innovation. The watch internals used surface-mounted components and flexible printed circuit boards. As well as the battery in the watch module, there was another located in the clasp, and connected to the radio by copper conductors inside the strap.

The clock section had a knob on the side which was twisted for On/Off, and three control buttons. Two of them were used to display and set the time, date and 24-hour alarm; the third was the backlight. The speaker section had a volume control on the side, and the tuner section had a tuning knob to select an FM frequency to listen to between 88 and 108 MHz.

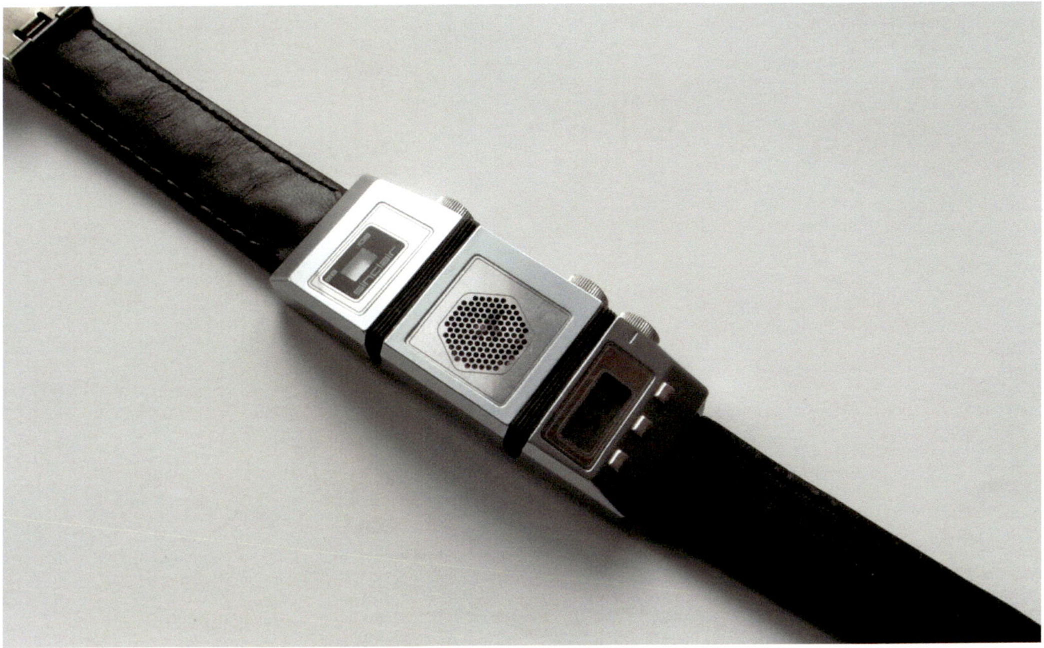

Sinclair's FM radio watch.

Hinged sections allowed the watch to be bent around the wrist.

The watch worked well, though the volume wasn't loud enough without it being held quite close to the ear. At its launch in 1985 at the Las Vegas Consumer Electronics Show it was marketed as the smallest FM radio watch in the world. Although it was well received at the show, unfortunately, because of financial difficulties and the acquisition by Amstrad, the watch never went into full production and only an estimated 11,000 were produced. The straps were made from a rubber compound which perished after a time and the thin copper wires inside the strap were extremely fragile and prone to breaking when flexed around the wrist. For these reasons, finding a working Sinclair FM Radio watch today would be a very rare find and be would likely cost a four-figure price.

Seiko T001 TV Watch (1982)

In the 1984 issue of the *Guinness Book of World Records*, Seiko's TV watch is listed as the smallest TV in the world. To gain that accolade, Seiko thought it was worth spending the millions of development dollars to produce a television that could be carried on the wrist. Seiko actually produced two earlier versions – the first for Tokyo and Osaka residents only and the second version that the whole of Japan could use. In 1983, Seiko brought out the T001-5010 for US customers at a price of $495 (more than $1,500 in today's money). However, you didn't actually get a TV set on your wrist. The wrist component comprised two screens – a small LCD at the top for the time and date and a 1¼-inch screen below for the monochrome video output. The other part was the TV receiver – a Sony Walkman-sized unit to be kept in the wearer's pocket or bag. The receiver allowed you to choose your favourite channel on either VHF (Very High Frequency), UHF (Ultra High Frequency) or FM radio. Having the separate receiver does make the 'smallest TV' claim slightly dubious but even so, the technology was way ahead of its time. The LCD screen comprised of

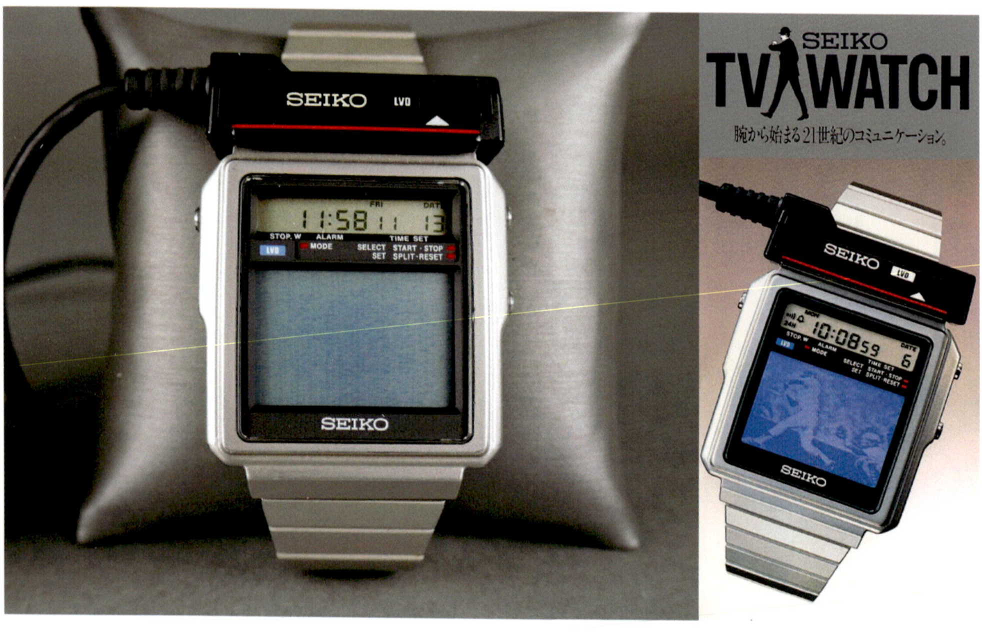

Seiko's TV watch was a technological marvel in 1984.

31,920 pixels each with ten shades of grey, though the picture quality was rather poor due to the slow response of early LCD screens, which gave rise to blurry motion. The batteries in the receiver would give five hours of use. The receiver attached to the watch via a six-conductor cable supplying three different voltages and separate signals for video and synchronisation. As well as TV, the watch had a stopwatch function and an alarm.

Despite the TV watch's practical drawbacks at the time, if you were to find one for sale today, you would be both very lucky and have to have deep pockets to own it. One of the reasons for its desirability is that it featured in the James Bond film *Octopussy* in 1983. Bond notices the Seiko TV watch in Q's lab just after he tested the tracking device on his Seiko G757 Sports watch (see earlier chapter). Bond asks Q what it is and Q replies: 'Oh, that's the latest Liquid Crystal TV.' Bond then zooms in on the bosom of one of the women working at the lab, and comments in 007-like chauvinistic manner: 'Perfect image, Q.' In the film, the watch is an early prototype, where the logo and the connector are different from the production one. The watch also has a colour screen, which unfortunately was not the case in reality. Apart from Bond, Tom Hanks can be seen wearing one in the comedy *Dragnet*.

Sadly, since analogue TV transmissions have been turned off for many years, these TVs can no longer be used. However, there are videos out there on the internet showing them receiving real TV programmes.

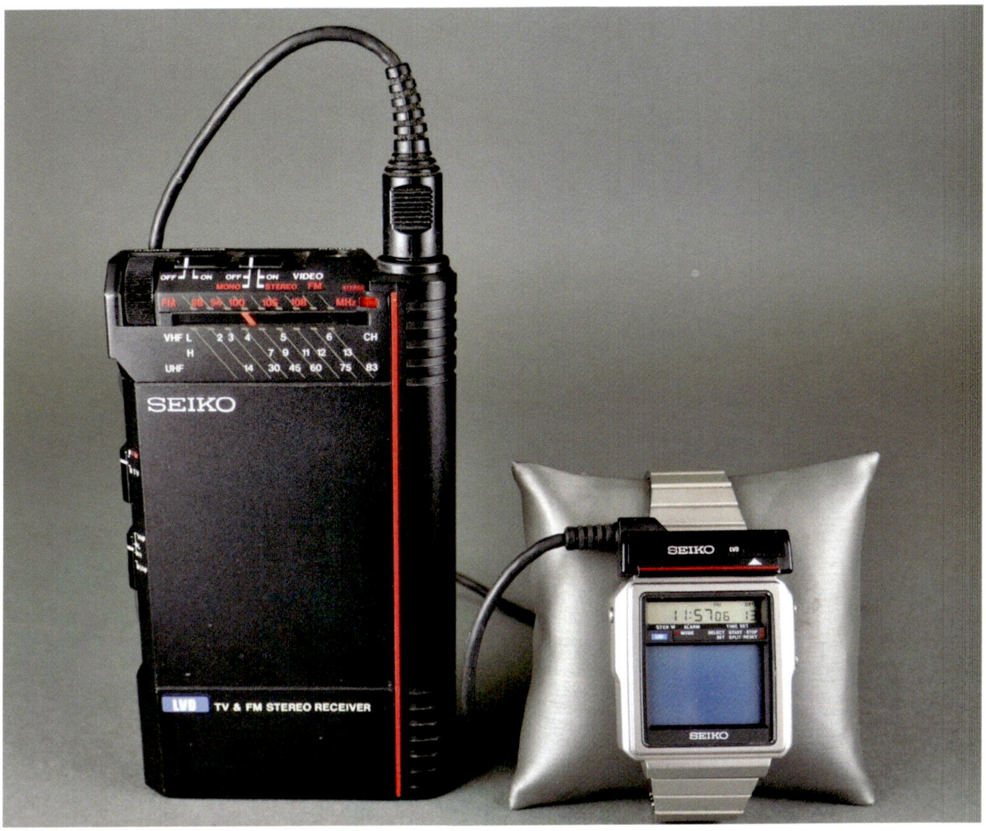

TV required connection to a large receiver.

Radio Shack 63-5042A Talking Watch (1985)

This 1985 watch from Radio Shack had the somewhat unique feature of speaking the time as well as via the more conventional LCD display. The Radio Shack company was started in 1921 by two brothers who wanted to provide radio equipment for the new field of amateur or ham radio. In 1939 they moved into the hi-fi market and sold their products under the Realistic brand, and in the early 1960s they expanded, adding eight more stores to the one in Boston. However, the business fell on hard times and was purchased in 1962 by Tandy Corporation, which at that time sold leather goods. Charles D. Tandy, the man behind the company, went on to build the largest chain of electronics stores in the world. Radio Shack stores tended to be small and cheap to run and were prolific throughout the US and the rest of the world. Eventually, Tandy and Radio Shack declined due to the fast-changing world of consumer electronics but they were a household name from the 1970s to the 1990s.

This watch's spoken time feature seems to have been marketed as a gimmick rather than a more useful aid for the visually impaired. The watch itself is not particularly attractive with its bulky monochrome appearance and large speaker dominating the front. Feature-wise, apart from the novelty of a synthesised voice, the watch had an alarm and an hourly chime.

Radio Shack's Talking Watch.

Casio CMD-40 Wrist Controller (1993)

This watch from Casio in 1993 combined the functions of stopwatch, calculator and TV controller in one neat package. Admittedly it was rather large at around 45 mm in length and nearly the same width, but it meant that losing the TV remote somewhere down the back of the sofa didn't matter anymore. The watch looks quite complex to use, but the instructions were user-friendly and once the watch had been set up to control your domestic appliance of choice, the rest was easy.

The watch has four modes which are cycled through via a button on the left-hand side: Timekeeping Mode (shown below), Calculator Mode, Alarm Mode and Stopwatch Mode. When in Timekeeping Mode, the left middle button cycles through four sub-modes: Normal Time Keeping, TV Mode (for television remote control), Video Mode (for video tape remote control) and Learning Mode. To set up the TV controller, the manual gave the code for your particular TV manufacturer (e.g. Samsung), which when entered into the watch gave you control. The same applied for your video recorder. If these didn't work or your manufacturer wasn't listed, you could try the Learning Mode, where the watch attempted to read the infrared signal from the remote control directly.

Casio's TV Remote Controller.

These watches never sold in big numbers, mainly because it was much easier to use the remote control for the job it was designed for, but they were a great gadget for the gadget collector.

Casio WQV-1 Wrist Camera (2000)

When Casio released the WQV-1 in Sept 2000, it was the world's first watch with a built-in digital camera. Costing around £149, it was cutting-edge tech and people were genuinely wowed that a wristwatch could take photographs. Remember that mobile phones with cameras built-in were still a few years away. This chunky watch was available in either a stainless steel or resin case and had a one-and-a-quarter-inch mono LCD screen on the front. The lens was positioned at the top, allowing you to point the camera by holding your arm out in front of you with the palm pointing downwards. The Casio-designed camera had a fixed F2.8 lens taking monochrome pictures and the 1 MB memory had room for 100 photos. The resolution was extremely poor by today's standards and the postage-sized photos were only 120 by 120 pixels, but hey, this was a watch with a built-in camera!

Taking photos was extremely easy. The shutter button on the front was pressed to put the watch in recording mode and the screen showed a live view. After framing the subject, pressing the shutter button again took the photo. Photos were stored with a time stamp, but text could be added later, such as a place name or a subject's phone number. To take the photos off the camera, there was an infrared port on the top next to the lens for connection to a PC. The watch came with an IR receiver port that plugged into a PC and suitable software

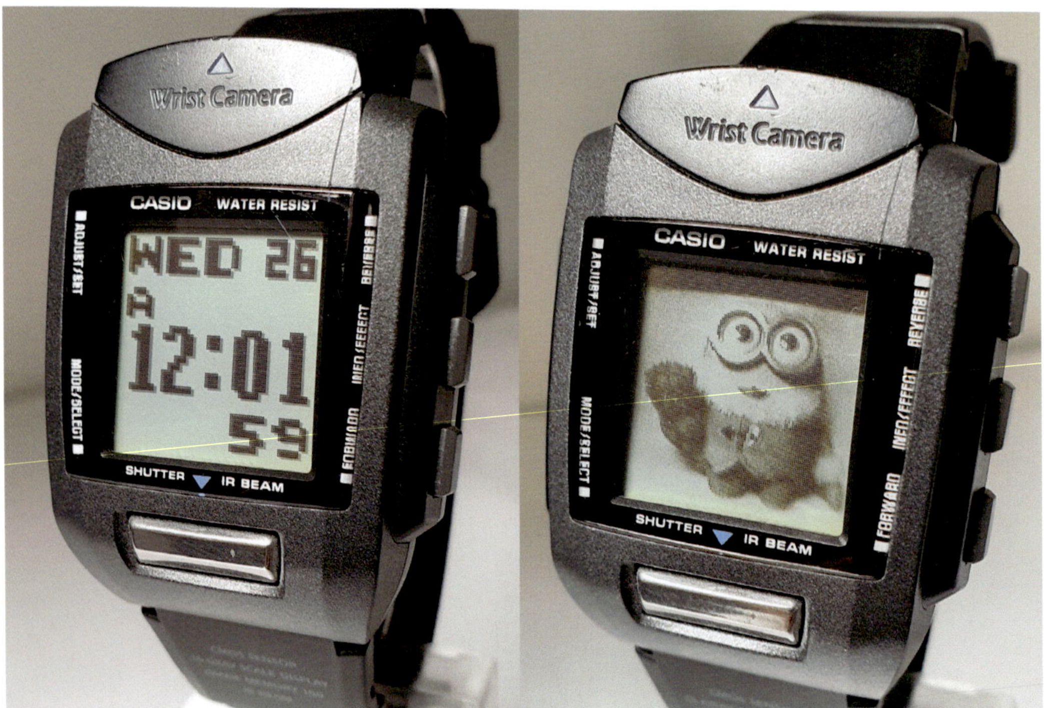

Casio WQV-1 was the first miniaturised digital camera.

to download the photos. As well as the camera, the watch also had five alarms, a sixty-minute countdown timer function and a stopwatch. Battery life was quoted at six months and the CR 2032 battery could be changed via a plastic coin-operated hatch in the rear. To save battery life the camera went into sleep mode after a couple of minutes. The WQV-1 was followed by the very similar WQV-2 and then a colour photo-capable WQV-3 model but with a mono screen. The final model was the WQV-10 released twelve months after the WQV-1.

Casio WQV-10 Wrist Camera (2001)

The WQV-10 was another first for Casio – the world's first colour camera watch with a colour screen. Although similar in physical size to its predecessor at approx. 40 mm by 50 mm, the design was improved to give the watch almost an art deco look. Apart from the colour screen, there were a few minor improvements over the WQV-1. The stored photos had a slightly higher resolution at 176 by 144 pixels and there was now a times-two digital zoom option but it was not much use in practice due to the low picture quality. To help with exposure, when in camera mode the bottom right button cycled through three lighting modes – indoor, outdoor and dusk. There was also a timer mode for the camera – either two, five or ten seconds - for those occasions when the watch owner wanted to be in the photo. Casio made full use of their new colour screen with colour icons instead of text where possible and there was even an option to change the colour 'theme' of the screen. Pressing the top right button for one second in visual data bank mode brought up the photo editing menu where you could delete a single photo or all of them, or add a name. Adding text was not a quick process though as you had to cycle through the alphabet to select each of the characters!

Colour camera was mounted on the top.

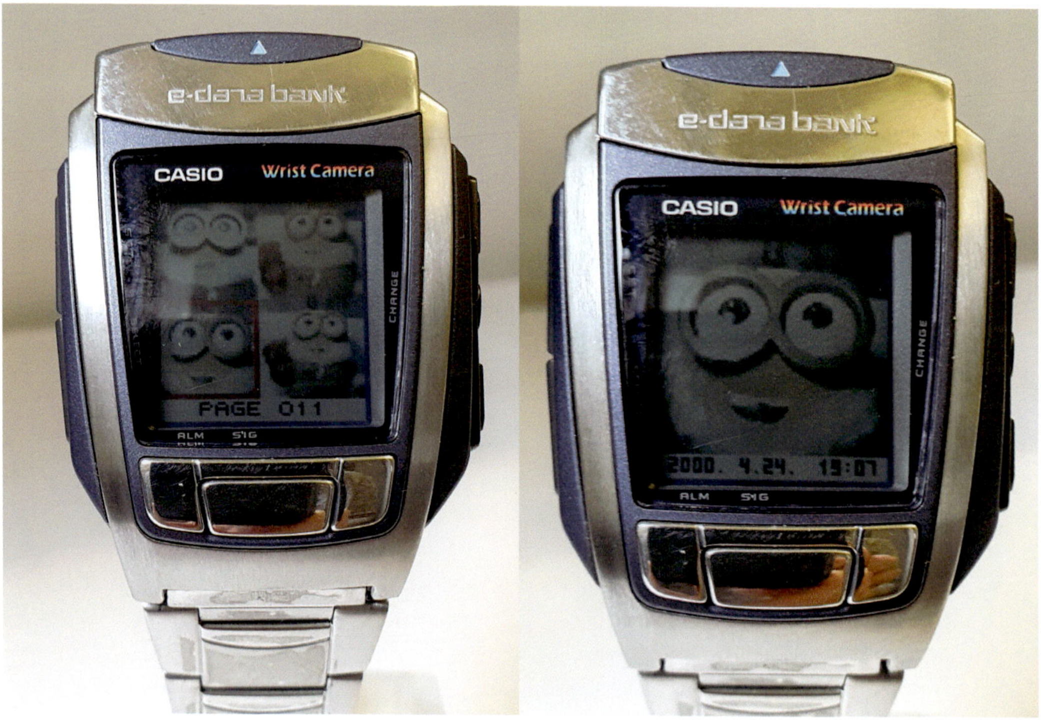

Tiny colour screen for viewing your photos.

Once happy with your photos though, the infrared comms screen offered different options for transmission to either another watch, a PDA, a computer or a game device. Playing with a WQV-10 is a great reminder of just hour far digital camera technology has come in the twenty-or-so years since their first appearance in a watch.

Polimaster PM1208

Although the PM1208 looks like a digital watch with some extra features, it is actually known as a dosimeter, or wearable radiation detector. Hidden under the face below the circular area at nine o'clock is a tiny Geiger Counter known as a Geiger-Muller tube, which measures gamma radiation. The tube requires a high voltage to operate and this is provided by a step-up coil and voltage multiplier connected to the conventional watch battery.

There are certain industries where workers can be exposed to radiation as part of the job, such as nuclear power plants or hospitals that use radioisotopes for treatments. Here, workers traditionally wear a film badge, which when processed reveals a wearer's total radiation exposure. The dosimeter is an electronic alternative for measuring and recording gamma radiation exposure over time.

Swiss company Polimaster has been making professional radiation detectors since 1992 and this model dates from the late 1990s. The dosimeter has two different measurement modes: DER (Dose Equivalent Rate) in micro-Sieverts per hour, and DE (Dose Equivalent) in milli-Sieverts. Before use, the wearer sets an appropriate radiation threshold for each mode and an alarm sounds if the threshold is exceeded. The DE value is also displayed

The Polimaster PM1208 radiation detector.

Somewhat larger than a conventional digital watch!

graphically in the circular analogue area and DER in the linear graphic area at the top. Up to 500 points of data are stored in non-volatile memory and this can be downloaded to a PC via a built-in infrared link. As well as dose measurement, the watch displays the time in both digital and analogue format; the hands can be independently set by pulling out the crown on the right-hand side. There is also an alarm and an electro-luminescent backlight.

These weighty devices are not your typical gadget watch, but are intended for professional use only. Hence, they are ruggedly made and designed for use in harsh environments where they might typically be worn over the top of protective rubber gloves. The large buttons on the sides give the wearer a fighting chance of operating the watch with a gloved hand.

Casio Smartwatches

Although not smartwatches in the modern sense of the word, this selection of watches from Casio demonstrates that many of the features that we think only became possible with modern smartwatches were in fact available years before.

1. Casio first produced a watch with a compass in 1993. There were a few versions of the CPW-100, but this model with the compass top-centre is a prayer compass, designed specifically to help Muslims face Mecca during prayers.

Casio watches with gadgets.

2. This watch with a UV sensor from 1994 gave an indication of how strong the sun was, together with a scale on the face showing maximum times in the sun for different skin types.
3. The VDB-1000 was an early PDA with built-in organiser, phone book, notepad and calculator. The touch-sensitive screen displayed ten by four characters, customisable by the watch's applications, plus five dedicated touch-sensitive buttons below the screen.
4. This novel watch from 1994, the VivCel VCL-100, had a built-in antenna to detect when your phone was ringing and notified you by vibrating on your wrist!
5. The world's first wrist-mounted MP3 player, the WMP-1 went on sale in the year 2000. Although the watch was rather large, its 32MB memory wasn't! It required a USB interface to a PC to load songs, and headphones attached to the watch at the nine o'clock position.
6. This watch in the G-Shock range, the GWS-9000, allowed wearers to make contactless payments in 2004 – ten years before Apple Pay. However, it was only accepted at a very limited number of places including Mobil filling stations and some McDonald's restaurants.
7. A watch with built-in GPS was available in 1999 with this PRT-1GP model. A GPS receiver at the time was quite sizeable, hence the bulge on the side incorporating the antenna.
8. Also from 1999, this model in Casio's data bank range, the DBC-V50 model, included a voice recorder function for recording notes when on the move.

More Casio watches with gadgets.

4

Smartwatches

Wikipedia defines the term 'smartwatch' as 'a wearable computer in the form of a watch'. The very first digital watches from Pulsar were actually marketed as 'Time Computers', due to the fact that they contained a microchip, and so they can claim to be the first smartwatches. But ever since, electronic watches have been evolving and adding more capability. In the intervening years, as we have seen, many watches appeared which had a unique feature added, such as a camera, radio or television.

Today, a smartwatch is usually used as an accessory to go with an owner's smartphone. Effectively, they become a second screen for your phone, sending you notifications, alerts, providing information such as a weather forecast, even making a phone call, all without having to remove your phone from your pocket. To do that, you need a true computer on your wrist, one that runs an Operating System (OS), onto which programs or apps can be loaded to perform specific tasks. Communication with the smartphone is normally by Bluetooth.

Smartwatches have taken a while to achieve this level of sophistication, but one of the people credited with starting smartwatch evolution was Canadian inventor Steve Mann. He designed the first watch running Linux in 1998, and for this he became known as the 'father of wearable computing'. But the era of the smartwatch that we would recognise today only started in the last fifteen years. This chapter describes some of the significant smartwatches from the twenty-teens.

Pebble (2012)

The Pebble smartwatch is credited by some as being the watch that created mainstream interest in the concept of linking a watch to a smartphone. It was unusual in that it wasn't developed by one of the big electronics giants, but was financed through crowdfunding. The American public benefit corporation Kickstarter managed to raise over $10 million in 2012, which was a record amount at the time. Investors in the project started receiving their watches less than a year later in January 2013, for the price of $99.

It could connect via Bluetooth to a tablet or phone running Android or Apple's IOS Operating System, and apps compatible with the Pebble could then be downloaded from an online app store as required. These included display of text notifications, emails and calls, activity tracking for steps and calories burned, remote control of compatible devices around the home, feeds from news apps and direction finding via GPS on the connected phone. In fact over 1,000 apps were available by 2014. Perhaps its most innovative feature was its display. Although only monochrome, this type of LCD display was known as 'transflective' and visible in all light conditions. In dim light, the display's transparency let the backlight shine through it and in bright conditions the display's reflective nature meant that it could be viewed even in direct sunlight, similar to a Kindle book reader. Charging was via a USB lead which attached magnetically and gave a seven-day battery life. This type of connection meant that the Pebble was also fully waterproof. Control of the watch was via four buttons, and there was also gesture detection via a built-in three-axis accelerometer. There was no sound but a motor gave vibration alerts.

The Pebble had a mixed reception when it was released. The good points were its design, readability and waterproofness. The negatives were a limit on the number of installed apps and lack of heart-rate monitor. These shortcomings were partly fixed in the second-generation Pebble Time models, which had a port on the back into which new hardware components like GPS and heart monitors could in theory be fitted. A Pebble 2 was released in 2016, but the company was wound up at the end of the year. But by then, over two million Pebbles had been sold.

Pebble 2 model from 2016.

Samsung Galaxy Gear (2013)

By 2013 all the major electronics companies were either developing or releasing their own smartwatches. One of these was Samsung who released the Galaxy Gear on 25 September of that year and was designed as a companion to Samsung's Galaxy range of smartphones. On paper, its specifications were impressive. The 1.6-inch square super AMOLED screen had a resolution of 320 by 320 pixels. Inside there was a 1.6 GHz ARM processor with 4GB of storage and 512MB of RAM, as well as an accelerometer and a gyroscope. Instead of a charging socket, the watch's 315 mAh battery was recharged via a micro-USB cradle, which also contained an NFC (Near Field Communication) tag for initial setup of the watch with the phone.

In the watch band there was also a camera for stills and video along with two microphones and a speaker. Communication with the host smartphone was with the recently introduced Bluetooth Low Energy (LE). LE is incompatible with traditional Bluetooth but gives improved battery life due to its low power. At the time this caused problems for customers as only Samsung devices which were running the very latest Android 4.3 supported Bluetooth LE. Smartwatch-to-phone communication was mainly used for displaying notifications from the phone on the watch, but the watch could also be used to make and receive phone calls via the dialler app on the watch.

Despite the amount of technology crammed into the watch, the Galaxy Gear faced a fair amount of criticism in initial reviews. The watch had limited compatibility due to specific versions of Android being required and the display of notifications was inconsistent. The components that were fitted into the strap made it uncomfortable to wear and the battery life was poor. Despite some pluses like the excellent camera and the ability to make phone calls directly, the Galaxy Gear was deemed a commercial failure, with less than 800,000 units manufactured.

Samsung Galaxy Gear SM-V700.

Sony Smartwatch 2 (2013)

Sony's contribution to the smartwatch market started in March 2012 with the imaginatively titled Sony Smartwatch, and at launch in the US it would have cost you around $125. The first thing to do once out of the box was to connect it to your Android phone via its Bluetooth connection. Then, you could go to the Google Play store and download tailor-made apps specifically made for Smartwatch, for capabilities such as display of SMS messages, weather reports, and Twitter. These apps could be simple 'widgets' giving you just one screen of information or full-blown applications. These were displayed on a 1.3-inch high-quality OLED display with 16-bit colour capability and 128 by 128 pixel resolution. Inside, the Smartwatch ran the Micrium operating system, an older OS used in industry to run on various microprocessors and pre-dates Android. It would connect to smartphones running Android version 4.0 or later. On the outside, the watch looks well-made with its lightweight aluminium body and choice of coloured silicone straps. However, there were a few minor annoyances. Firstly, the charging cable was temperamental and didn't always locate in the back of the watch or had a poor connection and so didn't charge. The touch sensitivity of the screen wasn't that great and tapping it twice to wake it up only worked if it was previously showing the clock. There was also no way to turn off the screen apart from waiting for it to time-out, and no brightness adjustment.

Sony Smartwatch 2 from 2013.

A year later Sony launched the Smartwatch 2, with a number of improvements. The screen was increased to 1.6 inches with resolution of 220 by 176 pixels, which meant the watch also got bigger. The OS was changed to use Google's Android Wear, which is a version of Android designed for smartwatches and wearables, and NFC communication was added. The final version, the Smartwatch 3, was released in November 2014 with further improvements such as a better processor, wi-fi connectivity and improved GPS which allowed the use of several walking, fitness and map-based apps to be used with it.

In conclusion, the Smartwatch from Sony was a more-than-competent device, but intense competition from other players in the market such as Apple and Google eventually forced Sony to withdraw.

Apple Watch (2015)

Arguably the most iconic and best known of all the smartwatches is the one from Apple. Apple would have liked to have called it the iWatch, in line with their other products such as the iPad and iPhone, but the iWatch trademark was already taken and so it was simply named the 'Apple Watch'. It was an immediate success, soon becoming the fastest-selling piece of wearable tech after its release in 2015. By 2020, over 100 million people were wearing one. Its success relied as much on Apple's genius marketing as its functionality. Apple had created a desirable object, and with its good looks and stylish design, it became as much a fashion accessory as a useful piece of technology.

The Apple Watch works in the same way as other smartwatches in that it runs an operating system, in this case, a version of watchOS, onto which apps can be installed.

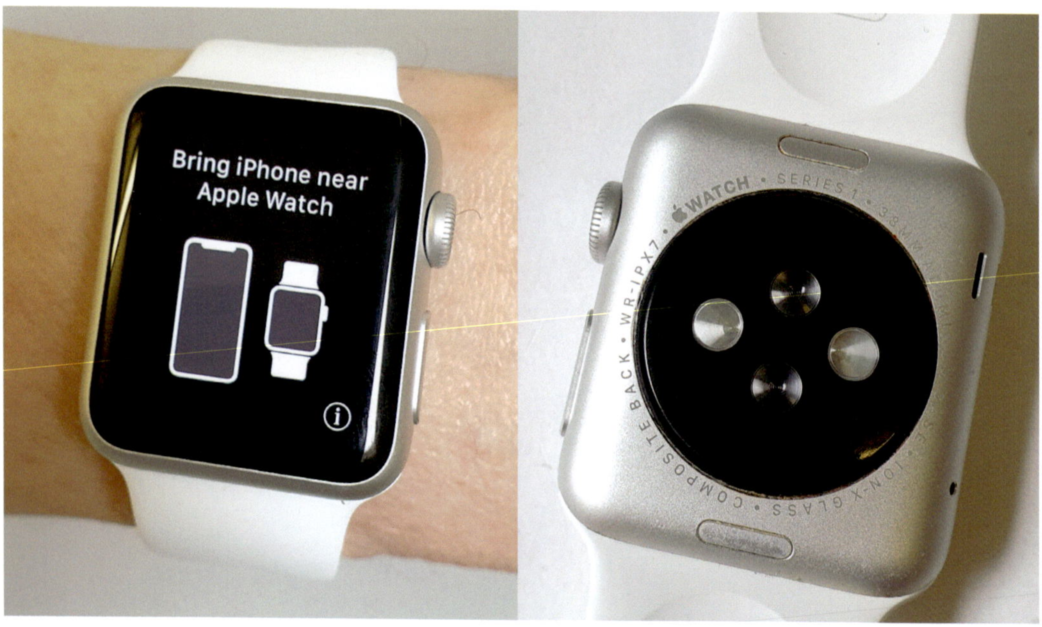

The original Apple Watch.

These apps can then send or receive data from the user's iPhone, for example texts or email notifications or uploading a route followed when exercising. In the absence of a phone, the watch can connect directly to a wi-fi network for internet access, and some models can also connect to a cellular network for making calls independent of an iPhone.

The first-generation watches (also known as Series 0) had 8GB of storage in total, of which 2 GB was for music and 75 MB for photos. Control was via a crown on the side which when turned, zoomed the screen or scrolled content. There was also a side button for displaying apps and for use in contactless payments. The 1.65-inch touch-sensitive OLED display had a 390 by 312 pixel resolution, and a touch feedback system known as a 'taptic engine'. This gave the screen a sort of pressure sensitivity and also vibrated to notify the wearer of an alert.

Since the initial release, Apple has launched a new model of its watch each September. These are known as 'series', and each series is available in many different variants. Case materials are either aluminium or stainless steel and Special Edition watches have titanium or ceramic cases. Some models are sold in conjunction with sponsors, e.g. Nike, and have branded bands, cases, special packaging and extras.

The watches contain a full complement of hardware sensors such as GPS (except Series 0), a gyroscope, accelerometer and barometer to determine orientation and altitude and the movements of the wearer. Apple has promoted its watch as a competitor to rivals' fitness trackers and as such, has the ability to monitor all the statistics from a workout. On the rear, green and infrared LEDs together with photodiodes are used to measure the reflected light from the skin, and from the difference in colour absorption, the heart rate can be calculated. Later models added electrical contacts on the back from which an ECG reading can be made. Models equipped with an additional red LED can also work out blood oxygen levels from the colour of the reflected light.

Apple watches have always been premium products with prices to match. The first watch would have cost around £500 in 2015, with current models going for over £800. If you must have the Apple accessory to go with your iPhone, then that probably seems like money well spent!

Series 4 Apple Watch (with protective cover).

5

Fitness Trackers

Although today's smartwatches have all of the functions you need to measure your physical activity, if you don't need anything extra, you could purchase a cheaper fitness tracker. They also tend to be more water resistant, as they are designed for swimming in addition to land activities. Watches to measure physical activity such as steps, heart rate and distance travelled have been around since the very early 1990s. They typically measured step count whilst walking or running and from that, worked out how far you had travelled and the number of calories burned. They became especially popular as a way of tracking the recommended 10,000 steps a day. Each step is measured using a built-in device called an accelerometer. Traditional accelerometers consisted of a mass attached to a spring, and a force acting on the mass (e.g. from a step) stretches the spring. Modern accelerometers have been miniaturised by either using the piezoelectric effect or by machining tiny springs made of silicon onto a chip. A software algorithm then converts the wearer's movements into a count of their steps. Early step counters required that your stride length be entered in order for total distance to be calculated, but in later models, built-in GPS gives you a much more accurate total distance travelled. Here are a couple of example fitness trackers from both ends of the price range.

Huawei Band 2 Pro (2017)

This model from Huawei in 2017 has plenty of features which made it good value for money. The watch itself is made of a combination of plastic and stainless steel, with a neat silicon rubber strap that fastens without a metal clasp to rust or get caught in clothing. The streamlined display fits within the wrist band profile and so makes it quite unobtrusive to wear, especially compared with later smartwatches with large circular faces. Control is via the button below the display, which is really just a touch-sensitive area. Tapping it cycles through the different activity functions (run, walk, cycle, swim) and flicking your wrist activates the normally off display. As well as a step counter, there is a heart-rate monitor and GPS to track your exercise. When your activity is over, you can sync to your phone and see all the statistics displayed in the Huawei app. And if

Above: Wristband-style Huawei Band 2 Pro.

Right: Data is displayed in the Huawei Health app.

that's not enough, the watch also has an oxygen monitor which calculates a measure of a person's aerobic capacity. One drawback with the GPS is that it sometimes takes a while to initialise and sometimes requests updates to its almanac for satellite positioning. On the plus side, the battery life is excellent for this sort of device at up to twenty-one days between charges. When released, the Band 2 Pro was around £50 but prices then fell to around £30 to keep it competitive.

However, Chinese-owned Huawei became mired in controversy after allegations were made that they used their telecoms equipment for the purposes of spying. This led to a ban on Huawei watch apps from being included in your phone's trusted app store. However, the app can be installed with a workaround so you don't have to go out and buy a Huawei phone.

Google Fitbit Versa 3 (2020)

Although the Versa 3 model from Fitbit is technically a smartwatch, Fitbit are perhaps *the* best-known name in fitness tracking technology and so it is included here. Founded in 2007 in the US, Fitbit is now a household name in wearable tech with over 120 million devices sold to date. Its range of fitness trackers started in 2009, and by 2014, Fitbit were offering trackers which could Bluetooth-sync with phones running either IOS or Android. Users could then log activity via the Fitbit app, and log steps, calories and their weight, and

Fitbit Versa 3 model from 2020.

set weekly goals and challenges. Since then they have gone from strength to strength and currently make a range of different wearable products.

Released in September 2020, the Fitbit Versa 3 model is part of their smartwatch range but is also a very competent fitness tracker. Its built-in GPS allows route data from, for example, a run or bike ride, to be logged along with heart rate and then stored and viewed on an Android phone after exercising. Although not the sleekest of designs at 12.35 mm thick, the Versa 3 is comfortable to wear with its soft silicone straps. Its charging lead attaches magnetically to the rear, a full charge lasting up to a week with light use. The watch is also water resistant to 50 m and so can monitor your stats whilst swimming. Apart from its fitness functions, the Versa 3 has a speaker and microphone included and so can be used for playing music or making phone calls.

Since 2021, Fitbit have been owned by Google's parent company Alphabet, and following privacy concerns that user's data might be used by Google without permission, Fitbit made a statement to customers assuring them that this would not happen. A positive from storing all this data has been that Fitbits have been used to convict criminals who would otherwise have escaped justice. In 2018 in the US, Anthony Aiello was convicted of murder after evidence from the victim's Fitbit showed that her heart rate had spiked then stopped shortly after a visit from Aiello.

Fitbit's versatile information screens.

Afterword

Even though smartwatches from the likes of Apple and Google are currently fashionable, their future is far from certain. Their small screen size limits how versatile they can be; for example, no one is going to watch a movie or a YouTube video on current-sized displays. But wearable tech also needs to be small and unobtrusive. And whilst battery life isn't great at the moment, this will undoubtedly improve over time. Many commentators predict that the next big revolution in wearables will be smart glasses, such as those offered by Google and others. Google Glass, for example, provides a head-up display experience by projecting data or video onto transparent screens which also double as lenses, the advantage being that you don't need to interrupt whatever you are doing to check your watch or phone screen – the information is literally right in front of your eyes. And as there is a voice-control option, you don't need to move your hands either. However, Google has ceased development on its Glass products and so only time will tell whether people will ditch their phones and watches for eye ware. Meanwhile, with a popular revival in old tech, many companies are remanufacturing retro hardware such as games consoles, record and cassette players and of course, digital watches, all for a new generation. Whatever new technology the future brings, I'm sure the humble electronic watch will be around for a long time to come in one form or other. Personally, I think I'll stick with my reliable old Casio.

Acknowledgements

Thanks go to the following people for granting permission to use their photos in the book: Lloyd Harvey at theledwatch.com for the Gold Pulsar P1, Dennis L. Klein at oldpulsars.com for the Pulsar P1 graphic and Michael Bolton at myretrowatches.com for the Commodore.

Also, thanks to all those at Amberley Publishing for continuing to allow me to collect and write about retro tech!